AUTOCAD ELECTRICAL 2015 BLACK BOOK

By

Gaurav Verma
CAD/CAM/CAE Expert
(IISD, Gurgaon)

Matt Weber
CAD/CAE Expert
(CADCAMCAE Works, Georgia)

Published by CADCAMCAE Works, USA. Copyright © 2014. All rights reserved. No part of this publication may be reproduced or distributed in any form or by any means, or stored in the database or retrieval system without the prior permission of CADCADCAE Works. To get the permissions, contact at cadcamcaeworks@gmail.com

ISBN # 978-1499505344

NOTICE TO THE READER

Publisher does not warrant or guarantee any of the products described in the text or perform any independent analysis in connection with any of the product information contained in the text. Publisher does not assume, and expressly disclaims, any obligation to obtain and include information other than that provided to it by the manufacturer.

The reader is expressly warned to consider and adopt all safety precautions that might be indicated by the activities herein and to avoid all potential hazards. By following the instructions contained herein, the reader willingly assumes all risks in connection with such instructions.

The Publisher makes no representation or warranties of any kind, including but not limited to, the warranties of fitness for a particular purpose or merchantability, nor are any such representations implied with respect to the material set forth herein, and the publisher takes no responsibility with respect to such material. The publisher shall not be liable for any special, consequential, or exemplary damages resulting, in whole or part, from the reader's use of, or reliance upon, this material.

DEDICATION

To teachers, who make it possible to disseminate knowledge
to enlighten the young and curious minds
of our future generations

To students, who are the future of the world

THANKS

To my friends and colleagues

To my family for their love and support

Training and Consultant Services

At CADCAMCAE WORKS, we provides effective and affordable one to one online training on various software packages in Computer Aided Design(CAD), Computer Aided Manufacturing(CAM), Computer Aided Engineering (CAE), Computer programming languages(C/C++, Java, .NET, Android, Javascript, HTML and so on). The training is delivered through remote access to your system and voice chat via Internet at any time, any place, and at any pace to individuals, groups, students of colleges/universities, and CAD/CAM/CAE training centers. The main features of this program are:

Training as per your need

Highly experienced Engineers and Technician conduct the classes on the software applications used in the industries. The methodology adopted to teach the software is totally practical based, so that the learner can adapt to the design and development industries in almost no time. The efforts are to make the training process cost effective and time saving while you have the comfort of your time and place, thereby relieving you from the hassles of traveling to training centers or rearranging your time table.

Current Software Packages on which we provide basic and advanced training are:

CAD/CAM/CAE: CATIA, Creo Parametric, Creo Direct, SolidWorks, Autodesk Inventor, Solid Edge, UG NX, AutoCAD, AutoCAD LT, EdgeCAM, MasterCAM, SolidCAM, DelCAM, BOBCAM, UG NX Manufacturing, UG Mold Wizard, UG Progressive Die, UG Die Design, SolidWorks Mold, Creo Manufacturing, Creo Expert Machinist, NX Nastran, Hypermesh, SolidWorks Simulation, Autodesk Simulation Mechanical, Creo Simulate, Gambit, ANSYS and many others.

Computer Programming Languages: C++, VB.NET, HTML, Android, Javascript and so on.

Game Designing: Unity.

Civil Engineering: AutoCAD MEP, Revit Structure, Revit Architecture, AutoCAD Map 3D and so on.

We also provide consultant services for Design and development on the above mentioned software packages

For more information you can mail us at:
cadcamcaeworks@gmail.com

Table of Contents

Training and Consultant Services   iv
About Author   xii

Chapter 1 Basics of Electrical Drawings

Need of Drawings	1-2
Electrical Drawings	1-2
Circuit Diagram	1-3
Wiring Diagram	1-4
Wiring Schedule	1-4
Block Diagram	1-5
Parts list	1-6
Symbols in Electrical Drawings	1-6
Conductors	1-6
Connectors and terminals	1-8
Inductors and transformers	1-9
Resistors	1-9
Capacitors	1-10
Fuses	1-11
Switch contacts	1-12
Switch types	1-13
Diodes and rectifiers	1-14
Earthing	1-15
Wire and Specifications	1-16
Types of Wires	1-16
Wire specifications	1-17
Labeling	1-18

Chapter 2 Introduction to AutoCAD Electrical and Interface

Introduction to AutoCAD Electrical	2-2
System requirements for AutoCAD Electrical 2015	2-2
Additional requirements for large datasets, point clouds, and 3D modeling	2-4
Starting AutoCAD Electrical Electrical	2-5
Creating a new drawing document	2-7
Meaning of Default templates	2-8
Title Bar	2-10
Changing Color Scheme	2-13
Application Menu	2-17
New options	2-18
Creating Drawings	2-19
Creating Sheet Sets	2-21
Open Options	2-25
Opening Drawing File	2-25
Save	2-27

Applying Password on File	2-29
Save As	2-30
Export	2-32
Publish	2-34
Print	2-34
Drawing Tab Bar	2-37
Drawing Area	2-38
Command Window	2-38
Bottom Bar	2-41
Drafting Settings dialog box	2-43

Chapter 3 Project Management

Workflow in AutoCAD Electrical	3-2
Initializing Project	3-3
Project Properties	3-8
Library Icon Menu Paths area	3-9
Catalog Lookup File Preference area	3-9
Options area	3-10
Opening a Project File	3-11
New Drawing in a Project	3-12
Refresh	3-15
Project Task List	3-15
Project Wide Update or Retag	3-15
Drawing List Display Configuration	3-18
Plotting and Publishing	3-19
Plot Project	3-20
Publish to WEB	3-22
Publish to DWF/PDF/DWFx	3-25
Zip Project	3-28
Removing, Replacing, and Renaming Drawings in a Project	3-30

Chapter 4 Inserting Components

Electrical Components	4-2
Inserting Component Using Icon Menu	4-2
Component Tag area	4-4
Catalog Data area	4-7
Description Area	4-10
Cross-Reference Area	4-11
Installation Code and Location Code	4-13
Pins area	4-14
Catalog Browser	4-14
User Defined List	4-18
Equipment List	4-20
Panel List	4-24
Pneumatic Components	4-25
Hydraulic Components	4-27
P&ID Components	4-27
Summary	4-28

Chapter 5 Wires, Circuits, and Ladders

Introduction	5-2
Wires	5-2
Wire	5-2
22.5 Degree, 45 Degree, and 67.5 Degree	5-4
Interconnect Components	5-5
Gap	5-5
Multiple Bus	5-6
Creating of Multiple Buses with various radio button	5-7
Ladders	5-10
Insert Ladder	5-11
XY Grid Setup	5-12
X Zones Setup	5-13
Wire Numbering	5-17
Wire Numbers	5-17
3 Phase	5-19
PLC I/O	5-20
Wire Number Leaders and Labels	5-21
Wire Number Leader	5-22
Wire Color/Gauge Labels	5-23
In-Line Wire Labels	5-25
Markers	5-26
Cable Markers	5-27
Multiple Cable Markers	5-29
Insert Dot Tee Markers	5-29
Insert Angled Tee Markers	5-30
Circuit Builder	5-31
Starting a New drawing	5-34
Editing Title Block	5-35
Creating Wires	5-36
Assigning Numbers and Labels to Wires	5-37
Inserting 3 Phase Motor	5-40
Adding Ground symbol	5-45
Adding symbols for various components	5-46

Chapter 6 Editing Wires, Components, and Circuits

Introduction	6-2
Edit Tool	6-2
Internal Jumper	6-4
Fix/UnFix Tag	6-8
Copy Catalog Assignment	6-8
User Table Data	6-10
Delete Component	6-12
Copy Component	6-13
Edit Circuits drop-down	6-13
Copying Circuit	6-14
Moving Circuit	6-15

Saving Circuit to Icon Menu	6-15
Transforming Components drop-down	**6-16**
Scooting	6-17
Aligning Components	6-17
Moving Component	6-18
Revering or Flipping Component	6-20
Re-tagging Components	**6-21**
Toggle NO/NC	**6-21**
Swap/Update Block	**6-22**
Swapping	6-22
Updating	6-23
Edit Attribute drop-down	**6-25**
Cross References Drop-down	**6-27**
Component Cross-Reference	6-28
Hide/Unhide Cross-Referencing	6-29
Update Stand-Alone Cross-Referencing	6-29
Circuit Clipboard panel	6-30
Editing Wires or Wire Numbers	**6-31**
Edit Wire Number	6-31
Fix	6-32
Swap	6-32
Find/Replace	6-33
Hide and Unhide	6-34
Trim Wire	6-34
Delete Wire Numbers	6-35
Move Wire Number	6-35
Add Rung	6-35
Revise Ladder	6-36
Renumber Ladder Reference	6-37
Wire Editing	6-38
Stretch Wire tool	6-39
Bend Wire tool	6-39
Show Wires	6-40
Check or Trace Wire tool	6-42
Wire Type Editing drop-down	6-43
Create/Edit Wire Type	6-43
Change/Convert Wire Type	6-45
Flip Wire Number	6-46
Toggle Wire Number In-Line	6-46

Chapter 7 PLCs and Components

Introduction	**7-2**
Inserting PLCs (Parametric)	**7-11**
Insert PLC (Full Units)	**7-13**
Addressing Area	7-15
Used Area	7-15
Tag	7-16
Options	7-17

Line1/Line2	7-17
Manufacturer	7-17
Catalog	7-17
Assembly	7-18
Catalog Lookup	7-18
Description	7-18
I/O Point Description Area	7-18
List descriptions	7-19
Pins	7-19
Show/Edit Miscellaneous	7-19
Ratings	7-19
Connectors	7-19
Insert Connector	7-20
Insert Connector (From List)	7-22
Insert Splice	7-23
Terminals	7-23
Inserting Terminals from Catalog Browser	7-23
Associate Terminals on the Same Drawing	7-24

Chapter 8 Practical and Practice

Introduction	8-2
Practical	8-2

Chapter 9 Panel Layout

Introduction	9-2
Icon Menu	9-4
Schematic List	9-9
Manual	9-11
Manufacturer Menu	9-12
Balloon	9-14
Wire Annotation	9-15
Panel Assembly	9-17
Editor	9-18
Table Generator	9-22
Insert Terminals	9-25
Editing Footprints	9-25
Edit	9-26
Copy Footprint	9-27
Delete Footprint	9-27
Resequence Item Numbers	9-29
Copy Codes drop-down	9-30
Copy Assembly	9-31
Page left blank intentionally for student notes	9-44

Chapter 10   Reports

Introduction	10-2
Reports (Schematic)	10-2

Bill of Materials reports	10-4
Component report	10-10
Wire From/To report	10-10
Component Wire List report	10-11
Connector Plug report	10-11
PLC I/O Address and Description report	10-11
PLC I/O Component Connection report	10-11
PLC Modules Used So Far report	10-12
Terminal Numbers report	10-12
Terminal Plan report	10-12
Connector Summary report	10-12
Connector Details report	10-12
Cable Summary report	10-12
Cable From/To report	10-13
Wire Label report	10-13
Missing Catalog Data	**10-13**
Electrical Audit	**10-15**
Drawing Audit	**10-17**
Dynamic Editing of Reports in Drawing	**10-18**
Modifying Tables	**10-19**
Modifying Rows	10-19
Modifying Columns	10-19
Merge Cells	10-20
Match Cells	10-20
Table Cell Styles	10-20
Edit Borders	10-20
Text Alignment	10-21
Locking	10-21
Data Format	10-22
Block	10-22
Field	10-23
Formula	10-23
Manage Cell Content	10-26
Link Cell	10-28
Download from source	10-28

Chapter 11   Project

Index   I-1

Preface

AutoCAD Electrical 2015 is an extension to AutoCAD package. Easy-to-use CAD-embedded electrical schematic and panel designing enable all designers and engineers to design most complex electrical schematics and panels. You can quickly and easily employ engineering techniques to optimize performance while you design, to cut down on costly prototypes, eliminate rework and delays, and save you time and development costs.

The **AutoCAD Electrical 2015 Black Book**, is written to help professionals as well as learners in performing various tedious jobs in Electrical control designing. The book follows a step by step methodology. This book is more concentrated on making you able to use tools at right places. The book covers almost all the information required by a learner to master the AutoCAD Electrical. The book starts with basics of Electrical Designing, goes through all the Electrical controls related tools and ends up with practical examples of electrical schematic and panel designing. Chapter on Reports makes you comfortable in creating and editing electrical component reports. Some of the salient features of this book are :

In-Depth explanation of concepts
Every new topic of this book starts with the explanation of the basic concepts. In this way, the user becomes capable of relating the things with real world.

Topics Covered
Every chapter starts with a list of topics being covered in that chapter. In this way, the user can easy find the topic of his/her interest easily.

Instruction through illustration
The instructions to perform any action are provided by maximum number of illustrations so that the user can perform the actions discussed in the book easily and effectively. There are about 1000 illustrations that make the learning process effective.

Tutorial point of view
The book explains the concepts through the tutorial to make the understanding of users firm and long lasting. Each chapter of the book has tutorials that are real world projects.

Project
Free projects and exercises are provided to students for practicing.

For Faculty
If you are a faculty member, then you can ask for video tutorials on any of the topic, exercise, tutorial, or concept.

Formatting Conventions Used in the Text
All the key terms like name of button, tool, drop-down etc. are kept bold.

Free Resources
Link to the resources used in this book are provided to the users via email. To get the resources, mail us at *cadcamcaeworks@gmail.com* with your contact information. With your contact record with us, you will be provided latest updates and informations regarding various technologies. The format to write us e-mail for resources is as follows:

Subject of E-mail as *Application for resources of _____ book*.
Also, given your information like
Name:
Course pursuing/Profession:
Contact Address:
E-mail ID:

For Any query or suggestion
If you have any query or suggestion please let us know by mailing us on *cadcamcaeworks@gmail.com*. Your valuable constructive suggestions will be incorporated in our books and your name will be addressed in special thanks area of our books.

About Authors

The author of this book, Gaurav Verma, has written many books on CAD/CAM/CAE books available already in market. His well-known books, **SolidWorks Simulation 2014 Black Book**, and **Creo Manufacturing 2.0 for Engineer and Machinists** are already available on Amazon. The author has hand on experience on almost all the CAD/CAM/CAE packages. Besides that he is a good person in his real life, helping nature for everyone. If you have any query/doubt in any CAD/CAM/CAE package, then you can contact the author by writing at cadcamcaeworks@gmail.com

The technical editor of the book, Matt Weber, has written books on various CAD packages. One of his best selling books, **SolidWorks 2014 Black Book** is available on Amazon for sale.

With the support of CreateSpace, we assure the best books and support to the education on latest technologies.

Basics of Electrical Drawings

Chapter 1

The major topics covered in this chapter are:

- *Need of Drawings*
- *Electrical Drawings*
- *Common Symbols in Electrical Drawings*
- *Wire and its Types*
- *Labeling*

NEED OF DRAWINGS

In this book, we are going through the topics related to electrical wiring and our purpose it to create good electrical wirings. So, it is important to know that why we need electrical drawings and what is the role of AutoCAD Electrical in that.

When we work in an electrical industry, we need to have a lot of information handy like the wiring of machines, position of switches, load of every machine and so on. It is almost impossible to remember all these details of an industry because there might be thousands of wires and switches and there can be hundreds of machine. To make this thing possible, we need electrical drawings that are written or printed documentation of these informations. Figure-1 shows an electrical drawing.

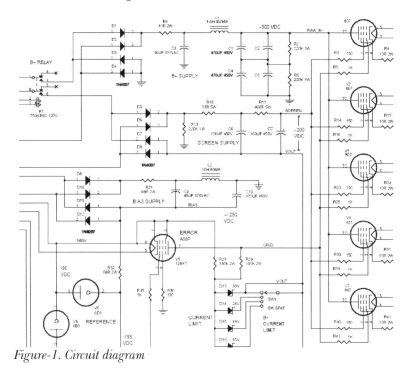

Figure-1. Circuit diagram

ELECTRICAL DRAWINGS

Electrical drawings are the representation of electrical components and connected wiring to fulfill a specific purpose. An electrical drawing can be of a house, industry or an electrical panel. Also, an electrical drawing can be divided into following categories:

- Circuit diagram
- Wiring diagram
- Wiring schedule
- Block diagram
- Parts list

Circuit Diagram

A circuit diagram shows how the electrical components are connected together and uses:

- Symbols to represent the components;
- Lines to represent the functional conductors or wires which connect them together.

A circuit drawing is derived from a block or functional diagram (see Figure-2). It does not generally bear any relationship to the physical shape, size or layout of the parts and although you could wire up an assembly from the information given in it, they are usually intended to show the detail of how an electrical circuit works.

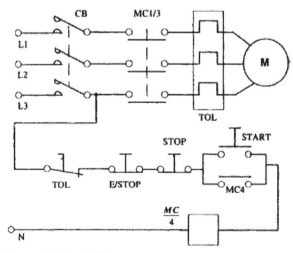

Figure-2. Circuit diagram

Wiring Diagram

This is the drawing which shows all the wiring between the parts, such as:
- Control or signal functions;
- Power supplies and earth connections;
- Termination of unused leads, contacts;
- Interconnection via terminal posts, blocks, plugs, sockets, lead-throughs.

It will have details, such as the terminal identification numbers which enable us to wire the unit together. Parts of the wiring diagram may simply be shown as blocks with no indication as to the electrical components inside. These are usually sub-assemblies made separately, i.e. pre-assembled circuits or modules. Figure-3 shows a wiring diagram.

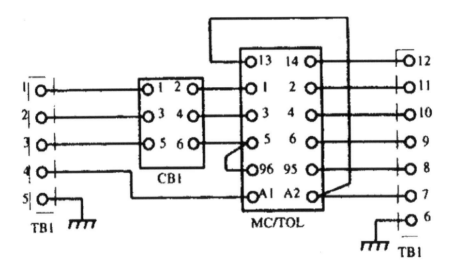

Figure-3. Wiring diagram

Wiring Schedule

A wiring schedule defines the wire reference number, type (size and number of conductors), length and the amount of insulation stripping required for soldering.

In complex equipment you may also find a table of interconnections which will give the starting and finishing reference points of each connection as well as other important information such as wire color, identification marking and so on. Refer to Figure-4.

Schedule:	Motor Control				206-A
Wire No	From	To	Type	Length	Strip Length
01	TB1/1	CB1/1	16/0.2	600 mm	12 mm
02	TB1/2	CB1/3	16/0.2	650 mm	12 mm
03	TB1/3	CB1/5	16/0.2	600 mm	12 mm
04	TB1/4	MC/A1	16/0.2	800 mm	12 mm
05	TB1/5	Ch/1	16/0.2	500 mm	12 mm

Figure-4. Wiring Schedule

Block Diagram

The block diagram is a functional drawing which is used to show and describe the main operating principles of the equipment and is usually drawn before the circuit diagram is started.

It will not give any real detail of the actual wiring connections or even the smaller components and so is only of limited interest to us in the wiring of control panels and equipment. Figure-5 shows a block diagram.

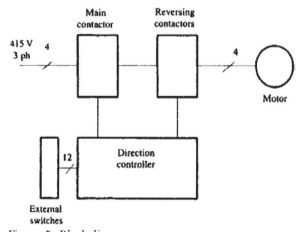

Figure-5. Block diagram

Parts list

Although not a drawing in itself, in fact it may be part of a drawing. The parts list gives vital information:

- It relates component types to circuit drawing reference numbers.
- It is used to locate and cross refer actual component code numbers to ensure you have the correct parts to commence a wiring job.

PARTS LIST			
REF	BIN	DESCRIPTION	CODE
CB1	A3	KM Circuit Breaker	PKZ 2/ZM-40-8
MC	A4	KM Contactor	DIL 2AM 415/50
TOL	A4	KM Overload Relay	Z 1-63

Figure-6. Parts list

You know various types of electrical drawings but these drawings contain various symbols. The following section explains the common symbols that are used in an electrical drawing.

SYMBOLS IN ELECTRICAL DRAWINGS

Symbols used in electrical drawings can be divided into various categories that are explained next.

Conductors

There are 12 types of symbols for conductors; refer to Figure-7 and Figure-8. These symbols are explained next.
1. General symbol, conductor or group of conductors.
2. Temporary connection or jumper.
3. Two conductors, single-line representation.
4. Two conductors, multi-line representation.
5. Single-line representation of n conductors.
6. Twisted conductors. (Twisted pair in this example.)

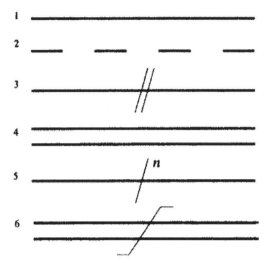

Figure-7. Symbols for conductors

7. General symbol denoting a cable.
8. Example: eight conductor (four pair) cable.
9. Crossing conductors - no connection.

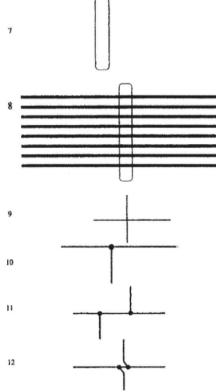

Figure-8. Symbols for conductors

10. Junction of conductors (connected).
11. Double junction of conductors.
12. Alternatively used double junction.

Connectors and terminals

Refer to Figure-9.
13. General symbol, terminal or tag.

These symbols are also used for contacts with moveable links. The open circle is used to represent easily separable contacts and a solid circle is used for those that are bolted.

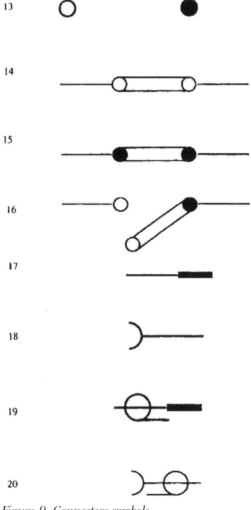

Figure-9. Connectors symbols

14. Link with two easily separable contacts.
15. Link with two bolted contacts.
16. Hinged link, normally open.
17. Plug (male contact).
18. Socket (female contact).
19. Coaxial plug.
20. Coaxial socket.

Inductors and transformers

Refer to Figure-10.
21. General symbol, coil or winding.
22. Coil with a ferromagnetic core.
23. Transformer symbols.

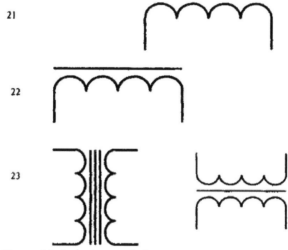

Figure-10. Inductors symbols

Resistors

Refer to Figure-11.
24. General symbol.
25. Old symbol sometimes used.
26. Fixed resistor with a fixed tapping.
27. General symbol, variable resistance (potentiometer).
28. Alternative (old).
29. Variable resistor with preset adjustment.
30. Two terminal variable resistance (rheostat).

31. Resistor with positive temperature coefficient (PTC thermistor).
32. Resistor with negative temperature coefficient (NTC thermistor).

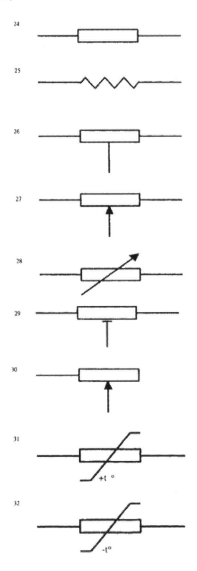

Figure-11. Resistors symbol

Capacitors

Refer to Figure-12.
33. General symbol, capacitor. (Connect either way round.)
34. Polarised capacitor. (Observe polarity when making

connection.)
35. Polarized capacitor, electrolytic.
36. Variable capacitor.
37. Preset variable.

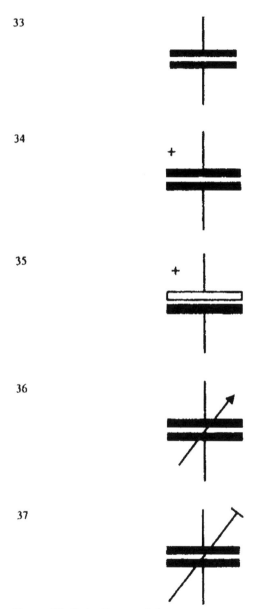

Figure-12. Capacitors symbols

Fuses

Refer to Figure-13.
38. General symbol, fuse.

39. Supply side may be indicated by thick line: observe orientation.
40. Alternative symbol (older).

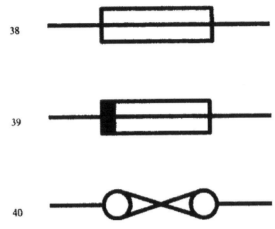

Figure-13. Fuses symbols

Switch contacts

Refer to Figure-14.
41. Break contact (BSI).
42. Alternative break contact version 1 (older).
43. Alternative break contact version 2.
44. Make contact (BSI).
45. Alternative make contact version 1.
46. Alternative make contact version 2.
47. Changeover contacts (BSI).
48. Alternative showing make-before-break.
49. Alternative showing break-before-make.

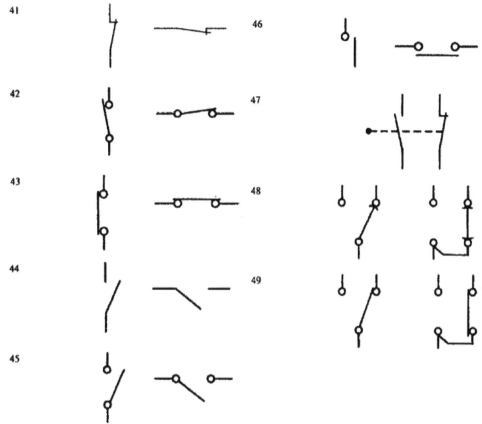

Figure-14. Switch Contact symbols

Switch types

Refer to Figure-15.
50. Push button switch momentary.
51. Push button, push on/push off (latching).
52. Lever switch, two position (on/off).
53. Key-operated switch.
54. Limit (position) switch.

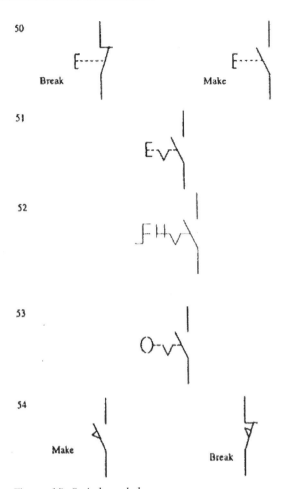

Figure-15. Switch symbols

Diodes and rectifiers

Refer to Figure-16.
55. Single diode. (Observe polarity.)
56. Single phase bridge rectifier.
57. Three-phase bridge rectifier arrangement.
58. Thyristor or silicon controlled rectifier (SCR) general symbol.
59. Thyristor - common usage.
60. Triac - a two-way thyristor.

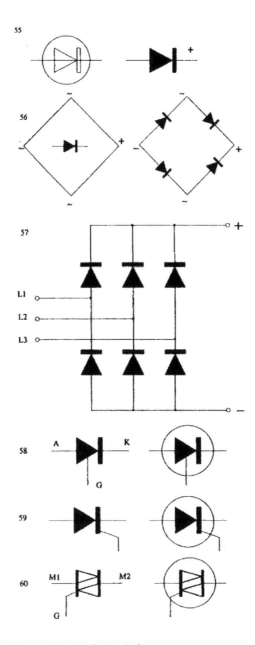

Figure-16. Diode Symbols

Earthing

Refer to Figure-17.

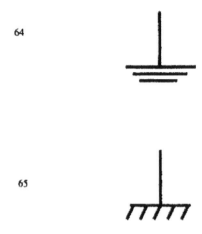

Figure-17. Earthing

Along with the above discussed symbols, you might need some user defined symbols for representation in your drawing.

After learning about various symbols the next important thing is to learn about wire and its specifications.

WIRE AND SPECIFICATIONS

Electrical equipment uses a wide variety of wire and cable types and it is up to us to be able to correctly identify and use the wires which have been specified. The wrong wire types will cause operational problems and could render the unit unsafe. Such factors include:

- The insulation material;
- The size of the conductor;
- What it's made of;
- Whether it's solid or stranded and flexible.

Types of Wires

- Solid or single-stranded wire is not very flexible and is used where rigid connections are accept able or preferred usually in high current applications in power switching contractors. It may be uninsulated.
- Stranded wire is flexible and most interconnections between components are made with it.

- Braided wire, also called Screened wire, is an ordinary insulated conductor surrounded by a conductive braiding. In this case the metal outer is not used to carry current but is normally connected to earth to provide an electrical shield to screen the internal conductors from outside electromagnetic interference.

Wire specifications

There are several ways to describe the wire type. The most used method is to specify the number of strands in the conductor, the diameter of the strands, the cross sectional area of the conductor then the insulation type.

Refer to Figure-18, Example 1:
- The 1 means that it is single conductor wire.
- The conductor is 0.6 mm in diameter and is insulated with PVC.
- The conductor has a cross-sectional area nominally of 0.28 mm .

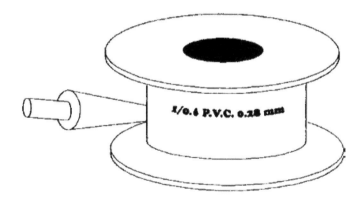

Figure-18. Example 1

Standard Wire Gauge

Solid wire can also be specified using the Standard Wire Gauge or SWG system. The SWG number is equivalent to a specific diameter of conductor; refer to Figure-19.
   For example; 30 SWG is 0.25 mm diameter.
   14 SWG is 2 mm in diameter.
   The larger the number – the smaller the size of the conductor.

There is also an American Wire Gauge (AWG) which uses the same principle, but the numbers and sizes do not correspond to those of SWG.

SWG table

SWG No.	Diameter
14 swg	2 mm
16 swg	1.63 mm
18 swg	1.22 mm
20 swg	0.91 mm
22 swg	0.75 mm
24 swg	0.56 mm
25 swg	0.5 mm
30 swg	0.25 mm

Figure-19. SWG table

We are at a position where we know about various schematic symbols and we know about wires. Now, we will learn about labeling of contactors.

LABELING

Labeling is the marking on components for identifying incoming and outgoing supply; refer to Figure-20. We also attach numbers to wires so that later on we can identify their circuits.

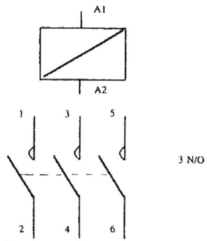

Figure-20. Contacts

Coils are marked alphanumerically, e.g. A1, A2.
  Odd numbers - incoming supply terminal.
  Even numbers - outgoing terminal.

Main contacts are marked with single numbers:
  Odd numbers - incoming supply terminal.
  Next even number - outgoing terminal.

In this way, we will find different type of markings for contacts that we would be including in our drawings.

Space Left for Notes

Space Left for Notes

Space Left for Notes

Introduction to AutoCAD Electrical and Interface

Chapter 2

The major topics covered in this chapter are:

- *Introduction*
- *System Requirement*
- *Starting AutoCAD Electrical/AutoCAD*
- *Starting Drawing*

INTRODUCTION TO AUTOCAD ELECTRICAL

In today's world, AutoCAD Electrical is a self defined name in the CAD industry. It is one of the beginners in Electrical CAD software industry. If we move back into the history, then first version of AutoCAD desktop applications came out around 1982 with the name AutoCAD Version 1.0. From 1982 to till today AutoCAD has undergone continuous enhancements and modifications. But, it is the speciality of AutoCAD that it still retains its position as No. 1 in CAD industry. The AutoCAD Electrical is built on that known platform called AutoCAD. The latest version AutoCAD Electrical 2015 is the most advanced model of AutoCAD Electrical available for us. The software has expanded more into the user interface of AutoCAD Electrical and has become the most user friendly one. For every tool/command it has more that one ways to invoke. This software also gives you the access to customize it as per your requirements.

Although the software is capable to perform 3D operations but in this book we will concentrate on 2D drawing creation. Now, we will learn to start AutoCAD Electrical and then we will discuss the interface of AutoCAD Electrical. But before we discuss about starting AutoCAD Electrical, Please check the system requirements to run AutoCAD Electrical 2015 properly. The system requirements are given next.

System requirements for AutoCAD Electrical 2015

Operating System
- Microsoft® Windows® 8/8.1
- Microsoft Windows 8/8.1 Pro
- Microsoft Windows 8/8.1 Enterprise
- Microsoft Windows 7 Enterprise
- Microsoft Windows 7 Ultimate
- Microsoft Windows 7 Professional
- Microsoft Windows 7 Home Premium

CPU Type

For 32-bit AutoCAD Electrical 2015:

32-bit Intel® Pentium® 4 or AMD Athlon™ Dual Core, 3.0 GHz or higher with SSE2 technology

For 64-bit AutoCAD Electrical 2015:

- AMD Athlon 64 with SSE2 technology
- AMD Opteron™ with SSE2 technology
- Intel® Xeon® with Intel EM64T support with SSE2 technology
- Intel Pentium 4 with Intel EM64T support with SSE2 technology

Network

- Deployment via Deployment Wizard.
- The license server and all workstations that will run applications dependent on network licensing must run TCP/IP protocol.
- Either Microsoft® or Novell TCP/IP protocol stacks are acceptable. Primary login on workstations may be Netware or Windows.
- In addition to operating systems supported for the application, the license server will run on the Windows Server® 2012, Windows Server 2012 R2, Windows Server 2008, Windows 2008 R2 Server editions, Windows Server 2003, and Windows 2003 R2 Server editions.
- Citrix® XenApp™ 6.5 FP1

Memory

2 GB (8 GB recommended)

Display Resolution

1024x768 (1600x1050 or higher recommended) with True Color.

Display Card

Windows display adapter capable of 1024x768 with True Color capabilities. DirectX® 9 or DirectX 11 compliant card recommended but not required.

Disk Space
Installation 6.0 GB

Pointing Device
MS-Mouse compliant device

Digitizer
WINTAB support

Plotter/Printer
Same as AutoCAD Electrical 2013-2014 — system printer and HDI support.

Media (DVD)
Download and installation from DVD

Browser
Windows Internet Explorer® 9.0 (or later)

Side-by-side Install
Supported

ToolClips Media Player
Adobe® Flash® Player v10 or up

.NET Framework
.NET Framework Version 4.5

Additional requirements for large datasets, point clouds, and 3D modeling

CPU Type
Intel Pentium 4 processor or AMD Athlon, 3.0 GHz or higher with SSE2 technology; Intel or AMD Dual Core processor, 2.0 GHz or higher.

Memory
8 GB RAM or greater

Disk Space

6 GB free hard disk available, not including installation requirements.

Display Card

1280x1024 True Color video display adapter 128 MB or greater, Pixel Shader 3.0 or greater, Direct3D®-capable workstation class graphics card.

Make sure that you fulfill all the requirements for the software before running it. Since there are various ways to perform the same thing so, we will use the best practice to perform an operation and alternates will be given

STARTING AUTOCAD ELECTRICAL ELECTRICAL

- Click on the **Start** button in the taskbar and type AutoCAD Electrical in the search box. The list of options related to AutoCAD Electrical will be displayed; refer to Figure-1.

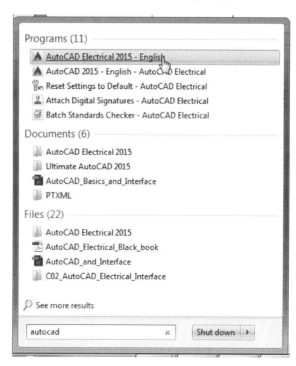

Figure-1. Start menu

- Click on AutoCAD Electrical link button at the top in the list. AutoCAD Electrical will initialize and after the processing the interface will be displayed as shown in Figure-2.

Figure-2. AutoCAD Electrical interface

The first screen of AutoCAD Electrical is divided into three columns. First column is for starting drawings. The second column is for opening the recent documents. Using the third column which connect, you can connect the other users of AutoCAD Electrical using your Autodesk account information.

Now, our first task is to create a drawing document. In real world we will find the drawings generally in two unit systems: Metric (SI) and Imperial. We will now learn to create drawings in both the unit systems.

CREATING A NEW DRAWING DOCUMENT

- Click on the down arrow below **Start Drawing** button in the First Column. List of drawing templates will be displayed; refer to Figure-3.

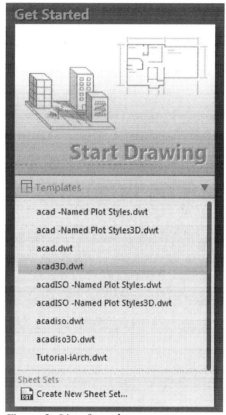

Figure-3. List of templates

- Select the desired template from the list. The drawing environment will open according to the selected template; refer to Figure-4.

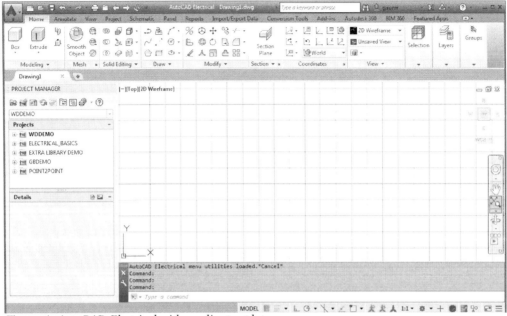

Figure-4. AutoCAD Electrical with acadiso template

Meaning of Default templates

acad -Named Plot Styles.dwt :- Using this template, we can create drawings in imperial unit system (feet and inches) that can be printed in black and white.

acad -Named Plot Styles3D.dwt :- Using this template, we can create 3D model in imperial unit system (feet and inches) that can be printed in black and white.

acad.dwt :-   Using this template, we can create drawings in imperial unit system (feet and inches) that can be printed in color.

acad3D.dwt :- Using this template, we can create 3D model in imperial unit system (feet and inches) that can be printed in color.

acadISO -Named Plot Styles.dwt   :- Using this template, we can create drawings in metric unit system (millimeters) that can be printed in black and white.

acadISO -Named Plot Styles3D.dwt   :- Using this template, we can create 3D model in metric unit system (millimeters) that can be printed in black and white.

acadiso.dwt :- Using this template, we can create drawings in metric unit system (millimeters) that can be printed in color.

acadiso3D.dwt :-   Using this template, we can create 3D model in metric unit system (millimeters) that can be printed in color.

Tutorial-iArch.dwt :-   Using this template, you can create architectural drawing that are compatible with the tutorial in default library. Note that the unit for this template is imperial (Feet and Inches).

Tutorial-iMfg.dwt :-   Using this template, you can create mechanical manufacturing drawing that are compatible with the tutorial in default library. Note that the unit for this template is imperial (Feet and Inches).

Tutorial-mArch.dwt :-   Using this template, you can create architectural drawing that are compatible with the tutorial in default library. Note that the unit for this template is metric (Millimeters.

Tutorial-mMfg.dwt :-   Using this template, you can create mechanical manufacturing drawing that are compatible with the tutorial in default library. Note that the unit for this template is metric (Millimeters).

No Template - Imperial   :-   Using this template, you can create a new drawing using the imperial unit system and without using any template.

No Template - Metric :- Using this template, you can create a new drawing using the metric unit system and without using any template.

Electrical Templates

The template files that start with ACE are meant for electrical drawings. Also, you can use ACAD Electrical.dwt and ACAD Electrical IEC.dwt for creating AutoCAD Electrical drawings. You will learn about these templates later in the book.

We know about the default templates. We have created a new document with a selected template. Now, we will learn about interface. We divide the AutoCAD Electrical interface into following sections:

- Title Bar
- Application Menu
- Ribbon
- Drawing Tab Bar
- Drawing Area
- Command Window
- Bottom Bar

Now, we will discuss each of the sections one by one.

TITLE BAR

Title bar is the top strip containing quick access tools, name of the document and connectivity options; refer to Figure-5.

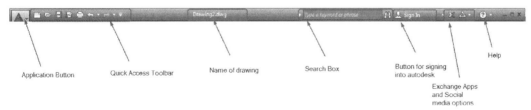

Figure-5. Title bar

- **Application** button is used to display **Application** menu, which we will discuss later.
- **Quick Access Toolbar** contains tools that are very common while working. The **Quick Access Toolbar** contains tools for

creating new file, saving current file, printing the file and so on. You can add the desired tools in the **Quick Access Toolbar** which we will learn later.
- The center of the title bar shows the name of the drawing.
- **Search Box** is used to search the desired topic in the **AutoCAD Electrical Help**. To use this option, type the keyword for which you want the information in the text box adjacent to search button and press **ENTER** from the keyboard.
- **Sign In** button is used to sign into the Autodesk account. If you are not having an Autodesk account then you can create one from Autodesk website. If you have an Autodesk account then you can save and share your files through Autodesk cloud and you can render on cloud which are very fast services. To sign into your Autodesk account, click on the **Sign In** button. As a result, a drop-down will display; refer to Figure-6. Click on **Sign In to Autodesk 360** button. As a result, the **Autodesk - Sign In** dialog box will be displayed; refer to Figure-7. Enter your ID and password and click on the **Sign In** button to login.

Figure-6. Sign In option

Figure-7. Autodesk sign in dialog box

- **Exchange Apps** is used to install or share apps for Autodesk products. Click on the **Exchange Apps** button and exchange apps web page will open in the browser where you can buy or try various apps as per your requirement.
- **Help** button is used to display the online help of AutoCAD Electrical on the Autodesk server. If you click on the down arrow next to help button then a list of options is displayed; refer to Figure-8. If you want to use offline help then click on the **Download Offline Help** option

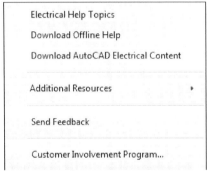

Figure-8. Help menu

Figure-9. Offline help link

Click on the desired language link to download the help file in respective language.

Now, we need to change the color scheme of AutoCAD Electrical because of printing compatibility although you can retain the present color scheme. To change the color scheme, steps are given next.

Changing Color Scheme

- Click on the **Application** button . The **Application Menu** will be displayed; refer to Figure-10.

Figure-10. Application menu

- Click on the **Options** button at the bottom right of the **Application Menu**. The **Options** dialog box will be displayed; refer to Figure-11.

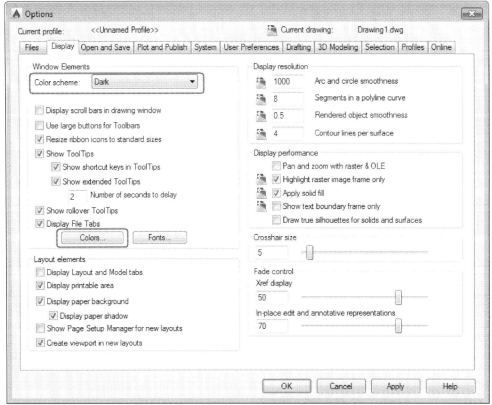

Figure-11. Options dialog box

- Click on the **Dark** drop-down button in the **Window Elements** area. List of options will be displayed.
- Select the **Light** option from the drop-down; refer to Figure-12.

Figure-12. Light option in Color scheme drop-down

- Click on the **Colors** button; refer to Figure-11. The **Drawing Window Colors** dialog box will be displayed as shown in Figure-13.

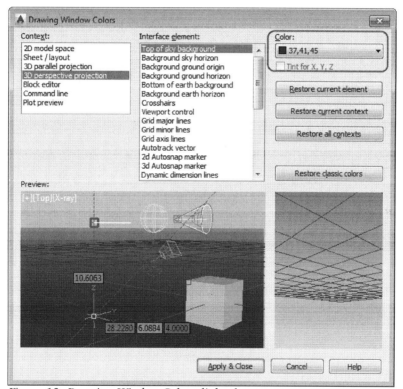
Figure-13. Drawing Window Colors dialog box

Note that using the options in this dialog box, you can change the color of any element of interface.

- Click on the drop-down for colors and select the White option; refer to Figure-14.

Figure-14. Color drop-down

- Similarly, one by one select the other background options from the **Interface element** box and set their colors to white.

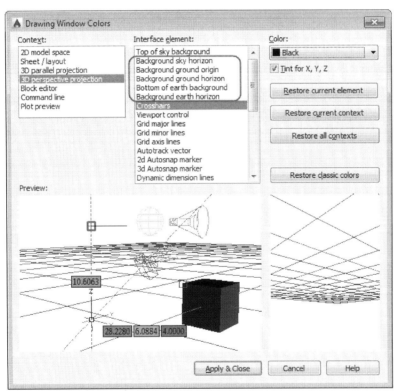

Figure-15. Background options

- Select the **Crosshairs** option from the **Interface element** box and select the **Black** option from the **Color** drop-down.
- Click on the **Apply & Close** button from the **Drawing Window Colors** dialog box and click on the **OK** button from the **Options** dialog box. The drawing area and interface will

be displayed in colors that are suitable for printing of this book.

APPLICATION MENU

Application menu contains the tools that are related to the overall functioning of AutoCAD Electrical. This functioning include (opening, saving, and creating of documents), (printing, publishing, and exporting of documents), and (Drawing properties and configuration options for AutoCAD Electrical). Now, one by one we will discuss each tool and option in the **Application Menu**.

New options

- Click on the **Application** button and hover the cursor over **New** option in the menu. The options for new documents will be displayed; refer to Figure-16.

Figure-16. Options for new documents

- In this version of AutoCAD Electrical there are two options to create new documents; **Drawing** and **Sheet Set**.

Using the **Drawing** option, you can create an individual drawing file, which contains model and various views of the model in orthographic projection.

Using the **Sheet Set** option, you can create a group of inter-related drawing files in the form of sheets.

Creating Drawings

- Click on the **Drawing** option from the **New** options. The Select Template dialog box will be displayed; refer to Figure-17.

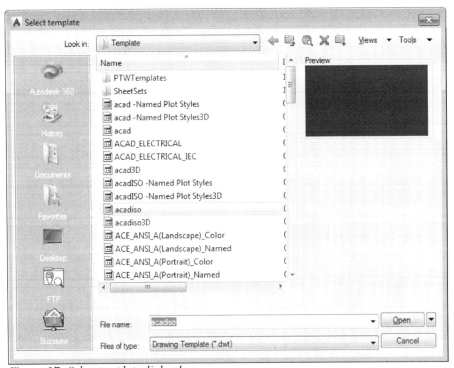

Figure-17. Select template dialog box

- Select the desired template from the dialog box. Note that the use of each template has already been discussed.
- If you want to create a document with out using any of the above template but with the desired units only then

click on the down arrow next to **Open** button in the dialog box. A drop-down will be displayed; refer to Figure-18.

Figure-18. Open drop down

- Click on the **Open with no Template - Imperial** button to work in Feet and Inches without loading any template.
- Otherwise, click on the **Open with no Template - Metric** button to work in millimeters without loading any template.
- Otherwise, click on the **Open** button from the drop-down to create a drawing using the selected template. Note that if you directly select the **Open** button from the dialog box then also the functioning will be similar.
- Note that from the **Files of type** drop-down in this dialog box the **Drawing Template (*.dwt)** option is selected so you are able to select the templates. You can also use any previous drawing as a template. To do so, click on the **Drawing Template (*.dwt)** option. The **Files of type** drop-down will expand as shown in Figure-19.

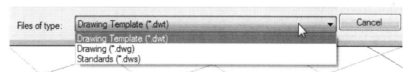

Figure-19. Files of type drop down

- Select the **Drawing (*.dwg)** option from the list and you will be able to select any drawing in place of template. Rest of the procedure is same.

Creating Sheet Sets

- Click on the **Sheet Set** option from the **New** options in the **Application Menu**. The **Begin** page of **Create Sheet Set** dialog box will be displayed; refer to Figure-20.

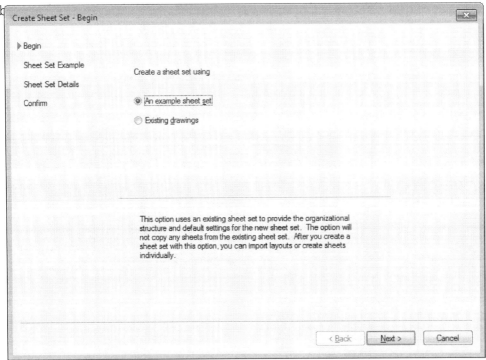

Figure-20. Begins page Create Sheet Set dialog box

- If you want to use the example sheet set (template) then select the **An example sheet set** radio button from this page. You can club more than one drawing files in the form of a sheet set by using the **Existing drawings** radio button. We will select the **An example sheet set** radio button for the time being.
- Click on the **Next** button from the dialog box. The **Sheet Set Example** page of **Create Sheet Set** dialog box will be displayed; refer to Figure-21.

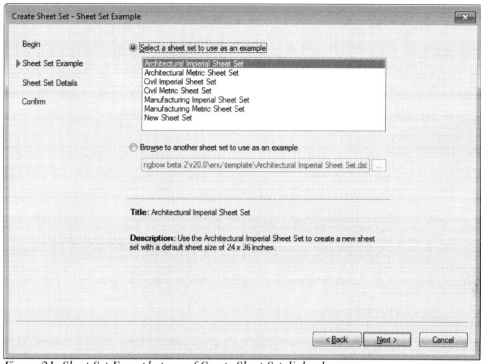

Figure-21. Sheet Set Example page of Create Sheet Set dialog box

- Select the desire sheet set template from the box and click on the **Next** button from the dialog box. The **Sheet Set Details** page will be displayed as shown in Figure-22.
- Click in the **Name of new sheet set** edit box and specify the name of the sheet set as desired.
- Click on the **Browse** button next to **Store sheet set data file (.dst) here** edit box. The **Browse for Sheet Set Folder** dialog box will be displayed as shown in Figure-23.

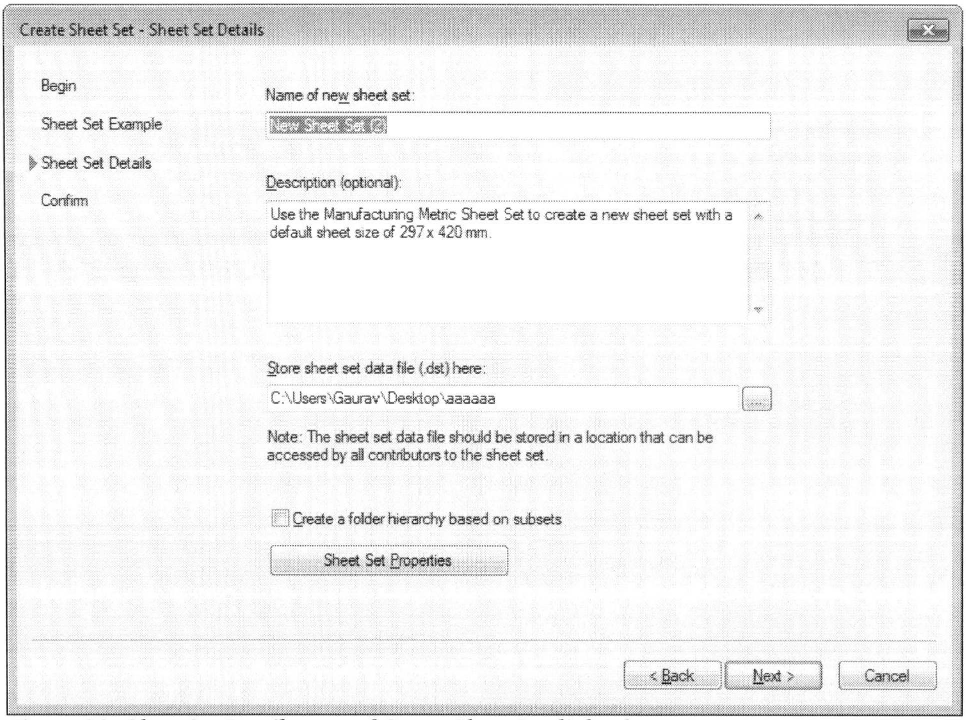

Figure-22. Sheet Set Details page of Create Sheet Set dialog box

Figure-23. Browse for Sheet Set Folder dialog box

- Browse to the desired folder and click on the **Open** button to set the directory.
- You can set the properties by using the dialog box displayed on clicking on the Sheet Set Properties button in this page.
- Click on the **Next** button from the page. The **Confirm** page will be displayed with the preview of all the properties set; refer to Figure-24.

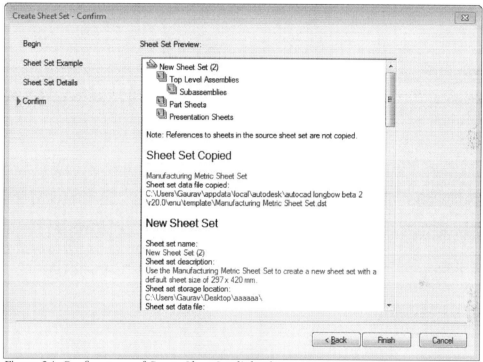

Figure-24. Confirm page of Create Sheet Set dialog box

- Click on the **Finish** button from the page. The **Sheet Set Manager** will be displayed with the current sheet set; refer to Figure-25.

Figure-25. Sheet Set Manager

We will learn more about the sheet sets later in this book.

Open Options

These options are used to open files of different formats and from different locations. In this version of AutoCAD Electrical, you can open drawing file from local drive or cloud (Autodesk 360), you can open sheet sets from local drive, you can import DGN format files, and you can open sample files from local drive as well as online. Note that the procedure to open all the type of files is similar, so here we will discuss the procedure to open a drawing file from local drive.

Opening Drawing File

- Click on the **Drawing** option from the **Open** options in the **Application Menu**. The Select File dialog box will be displayed as shown in Figure-26.

- Browse to the desired folder and select the file that you want to open in AutoCAD Electrical.
- Click on the **Open** button to open the file.

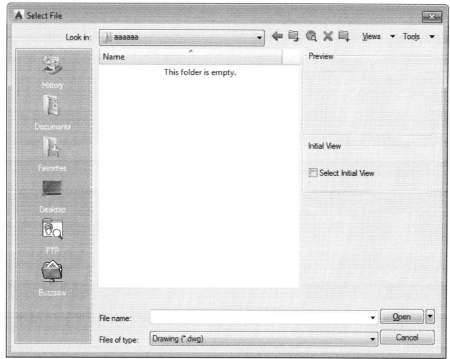
Figure-26. Select File dialog box

Note that there are three more methods to open a file; opening as read only, opening partially, and opening as read only with partial content. The options to do so are displayed on clicking at the down arrow next to **Open** button; refer to Figure-27. Partially opening a drawing file mean skipping some of the layers that are not of our use. A layer is transparent sheet on which we draw something in AutoCAD Electrical. In AutoCAD Electrical, each drawing is stacking of multiple layers one over the other.

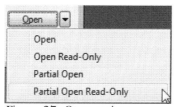
Figure-27. Open options

- If you want to open the file as read only and don't want to make the changes in it then select the **Open Read-Only** option from the drop-down.
- If you want to open a drawing file partially, then select the **Partial Open** option from the drop-down. The **Partial Open** dialog box will be displayed as shown in Figure-28.

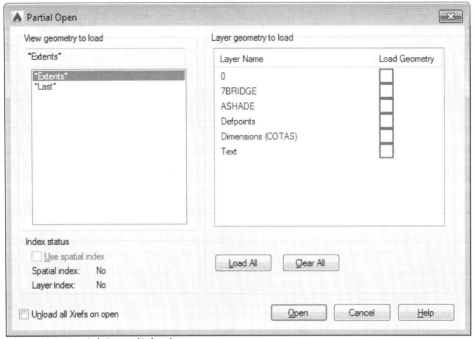

Figure-28. Partial Open dialog box

- Select the check boxes corresponding to the layers that you want to include while opening the drawing.
- Click on the **Open** button from the dialog box. The file will open with only the selected layers.

You can similarly open a file partially and read only by using the **Partial Open Read-Only** option.

Save

This option is used to save the current file in the local drive for its later use. The steps to save the file are given next.

- Click on the **Save** option from the **Application Menu**. The Save Drawing As dialog box will be displayed as shown in Figure-29.

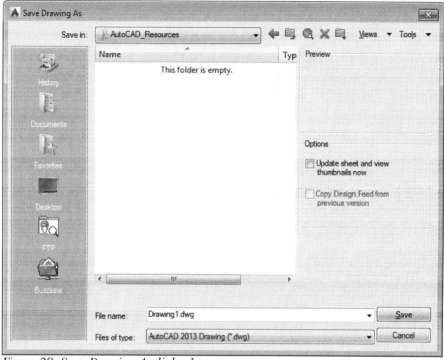

Figure-29. Save Drawing As dialog box

- Change the file name as desired and click in the Files of type drop-down to change the file format. On clicking at the **Files of type** drop-down, the list of formats is displayed as shown in Figure-30.

Figure-30. AutoCAD Electrical file formats

- Browse to the desired location where you want to save the file. Note that the buttons in the left area of the

dialog box are used to access some common locations like Documents, Desktop, FTP, and so on.
- Click on the **Save** button and the file will be saved with the specified settings.

Applying Password on File

Note that you can apply a password to the drawing file for security of your design. To apply the password follow the steps given next.

- In the **Save Drawing As** dialog box, click on the down arrow next to **Tool** button; refer to Figure-31. A list of options will be displayed.

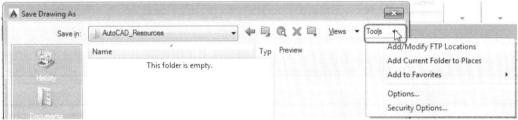

Figure-31. Tools drop down

- Click on the Security Options button in the drop-down. The Security Options dialog box will be displayed as shown in Figure-32.

Figure-32. Security Options dialog box

- Specify the password for security in the edit box corresponding to password.
- Select the **Encrypt drawing properties** check box to encrypt the properties of the drawing.
- Note that after specifying the password, the **Advanced Options** button becomes active. Click on this button and select the encryption provider as per your requirement. If you are sending this file to someone else then make sure that the person has the same encryption agent in this system.
- Click on the **OK** button from the dialog box. The **Confirm Password** dialog box will be displayed as shown in Figure-33.

Figure-33. Confirm Password dialog box

- Type your password again and click on the **OK** button from the dialog box. Next, click on the **Save** button from the dialog box.

Save As

This option is used to save the file with a different format. The steps to use this option are given next.

- Click on the arrow next to **Save As** in the **Application Menu**. The **Save As** options will be displayed as shown in Figure-34.

Figure-34. Save As options

- If you want to save the file with a format different from the AutoCAD Electrical format then click on the **Other Formats** button from the list of options. The **Save Drawing As** dialog box will be displayed as discussed earlier. Note that some additional formats will be displayed in the **Files of type** drop-down.
- Select the desired file format from the list and click on the **Save** button from the dialog box to save the file. Note that later when we will work on 3D models and some more options will be available in this drop-down.

Export

The options for this section are used to save the file in external formats. The options in this section work in the same way as the **Save** options work. The steps to export a file in iges format are given next. You can apply the same procedure to save files in other formats.

- Hold the cursor on the **Export** option in the **Application Menu**. The options related to exporting file are displayed; refer to Figure-35.

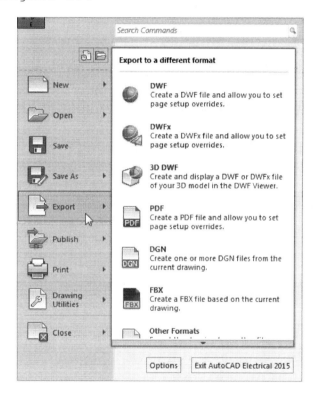

Figure-35. Export options

- Click on the **Other Formats** button from the list. The **Export Data** dialog box will be displayed as shown in Figure-36.

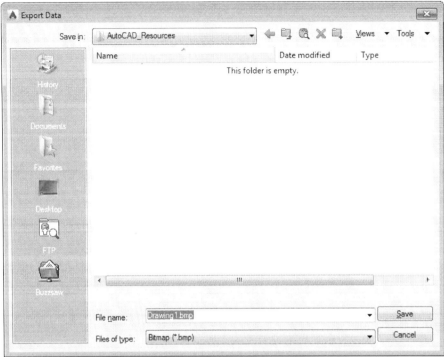

Figure-36. Export Data dialog box

- Specify the desired name in the File name edit box and click on the Files of type drop-down. The list of formats will be displayed as shown in Figure-37.

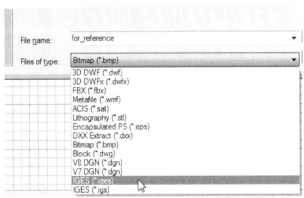

Figure-37. Formats for export

- Select the **IGES (*.iges)** option from the list and click on the **Save** button from the dialog box to save the file with the selected format.

Publish

The options for this section are used to package and transfer the file to external sources like 3D printer, web services and e-mail transfer.

- To use the **Send to 3D Print Service** option, you must have a 3D model which can be sent for 3D printing. If you have the model then click on this button and select the models that you want to print.
- **Archive** option is used to package the files for transfer. Note that to use this option, you need to create a sheet set of drawing. If you have the sheet set then click on this button and you can package the files in a zip format.
- **eTransmit** option is also used to package the files but it compresses the files to their fast electronic transfer.
- **Email** option is used to directly e-mail the current file to your client.

Print

Printing is an important requirement of the CAD industry. All the designs are manufactured and controlled on the basis of printed copy of the drawing. To get the print out follow the steps given next.

- Click on the **Plot** option from the **Print** options in the Application Manager. The **Plot** dialog box will be displayed for Model option; refer to Figure-38. **(Model is the workspace in AutoCAD Electrical in which 3D view of model is displayed. Layout/Paperspace is the workspace where model is** displayed in its paper print form. It is advised to shift to Layout/Paperspace before printing on the paper.).

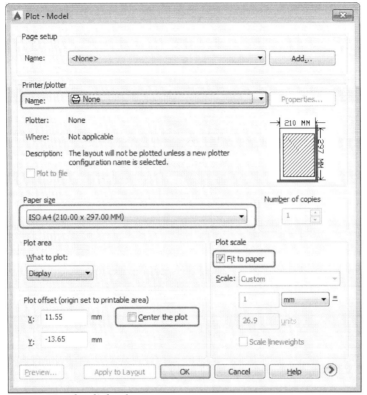

Figure-38. Plot dialog box

- Select the Printer/plotter from the **Name** drop-down in the **Printer/plotter** area of the dialog box; refer to Figure-39.

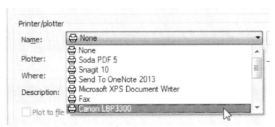

Figure-39. Printer selection

- Select the paper size from the drop-down in the **Paper size** area of the dialog box.
- Select the **Center the plot** check box to print the drawing at the center of the paper.
- Select the **Fit to paper** check box to fit it in the current paper size.

- Click on **More options** button or press **ALT + >** to expand the dialog box. The dialog box will be displayed as shown in Figure-40.

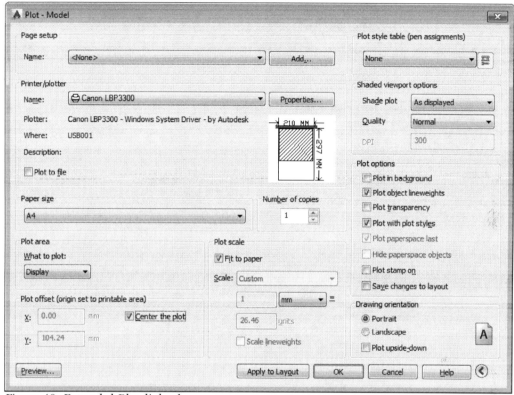

Figure-40. Expanded Plot dialog box

- Select the **Landscape** radio button from the **Drawing orientation** area of the dialog box to print the drawing in landscape orientation.
- Select the **Plot upside-down** check box from the **Drawing Orientation** area to reverse the printing side of the paper.
- To increase or decrease the quality of printout, select the desired quality option from the **Quality** drop-down in the **Shaded viewport options** area.
- Click on the **OK** button from the dialog box to print the drawing. You will learn more about the printing later in this book.

DRAWING TAB BAR

The **Drawing tab bar** contains tabs for each drawing that is opened; refer to Figure-41.

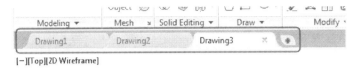

Figure-41. Drawing Tab bar

There are various functions that can be performed by using the Drawing tab bar, which are discussed next.

- Hold the cursor over the tile of any drawing file in the Drawing tab bar. An interactive box will be displayed allowing you to switch between model and their paper spaces/layouts; refer to Figure-42.

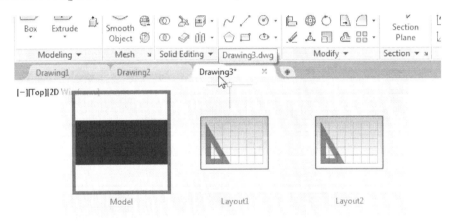

Figure-42. Interactive box for layout switching

- Click on the **Layout1** or **Layout2** to switch to paper space mode. You can switch back to Model space mode by using the same procedure.

- You can create a new drawing file by clicking on the plus sign next to the drawing tile. On doing so, the Start page of AutoCAD Electrical will be displayed.
- Using the options in this page, you can start a new drawing as discussed earlier.

DRAWING AREA

The blank area in below **Drawing Tab bar** is called **Drawing Area**. This area is used to create sketches and models. Figure-43 shows the annotated drawing area. You will learn about more options related to drawing area later in this book.

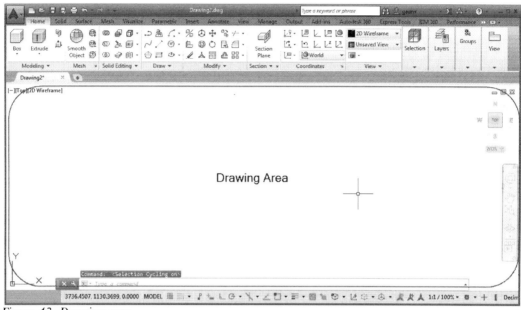
Figure-43. Drawing area

COMMAND WINDOW

This window is used to start new commands or specify the parameters for current running commands. This window is available above the **Bottom bar**; refer to Figure-44.

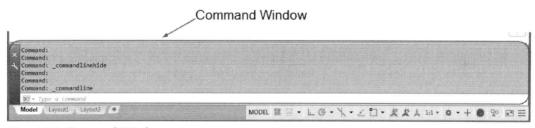
Figure-44. Command Window

You can press **CTRL+9** to display/hide the command window. The methods to use the command window are given next.

- Click in the **Type a Command** box in the Command window. You are prompted to specify the command.
- Type a few alphabets of your desired command, a list of commands relevant to your specified alphabets will be displayed; refer to Figure-45.

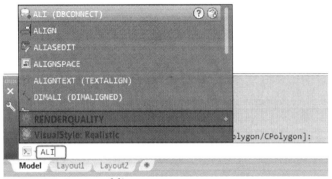

Figure-45. Command list

- Hold the cursor on the name of a command, the description of the command will be displayed; refer to Figure-46.

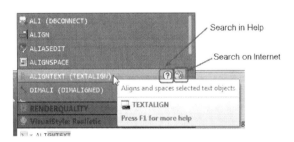

Figure-46. Tool description

- To know more about the highlighted command click on the **Search in Help** or **Search on Internet** button next to the command name.
- You can scroll down in the box to browse the command not visible currently in the box. You can also specify the values of various variables.
- Click on the **+** sign next to the variables tile in the box; refer to Figure-47. The list of variables related to the specified alphabets will be displayed; refer to Figure-48.

Figure-47. Variables

Figure-48. List of variables

- To start any command or specify value of any variable click on it in the box. The respective options will be displayed in the command window and you will be prompted to specify the desired values. For example, type **L** at the command prompt in the **Command Window** and click on the **Line** tool at the top in the command box. You will be prompted to specify the position of the starting point of the line; refer to Figure-49. You will learn more later in this book. Press **ESC** to exit the tool.

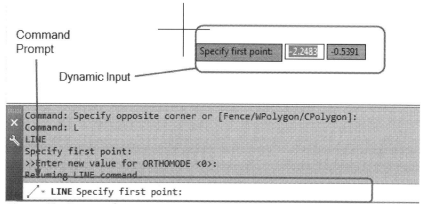
Figure-49. Command prompt

Bottom Bar

The tools in the **Bottom Bar** are used to perform various functions like creating a new layout or switching between existing ones, enabling orthogonal movement of cursor and so on; refer to Figure-50. The options in the Bottom Bar are discussed next.

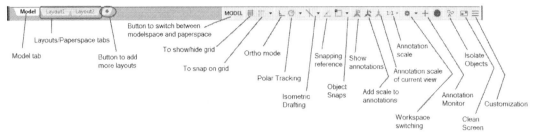

Figure-50. Bottom bar

Model tab and Layout tabs

Click on these tabs to switch between model space and paper space. To add more layouts, click on the **+** sign next to the Layout tabs.

In a layout, you can switch between paper space and model space by using the **MODEL** button in the Bottom Bar.

Grid Display

In AutoCAD Electrical, the grid lines are used as reference lines to draw objects. If the **Grid Display** button is chosen, the grid display is on and the grid lines are displayed on the screen. The F7 function key can be used to turn the grid display on or off.

Grid Snap

This option is used to enable snapping of cursor to the grid points as per the specified settings. Snapping is the attraction of cursor to the key points of AutoCAD Electrical. There are two options for specifying grid snap: Grid Snap button and corresponding drop-down. The steps to use Grid Snap are given next.

- Click on the **Grid Snap** button to enable snapping of cursor to the key points on grid.
- Click on the down arrow next to **Grid Snap** button; a flyout will be displayed as shown in Figure-51.

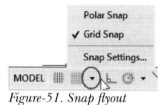

Figure-51. Snap flyout

- Select the **Grid Snap** option if you want to snap the cursor to the grid intersections. Select the **Polar Snap** option from the flyout if you want to snap the cursor to the angular lines.
- Select the **Snap Settings** option from the flyout to specify the setting related to the selected type of Snap. The **Drafting Setting** dialog box will be displayed with the **Snap and Grid** tab selected; refer to Figure-52.

Figure-52. Drafting Settings dialog box

- After specifying the desired setting click on the OK button from the dialog box. The cursor will automatically start snapping to the key points as specified by the settings. Various options in the Drafting Setting dialog box are discussed next.

Drafting Settings dialog box

You can use the **Drafting Settings** dialog box to set the drawing modes such as grid, snap, Object Snap, Polar, Object snap tracking, and Dynamic Input. All these modes help you draw accurately and also increase the drawing speed. You can right-click on the Snap Mode, Grid Display, Ortho Mode, Polar Tracking, Object Snap, 3D Object Snap, Object Snap Tracking, Dynamic Input, Quick Properties, or Selection Cycling button in the Status Bar to display a shortcut menu. In this shortcut menu, choose Settings to display the Drafting Settings dialog box, as shown in Figure-52. This dialog box has seven tabs: **Snap and Grid**, **Polar Tracking**, **Object Snap**, **3D Object Snap**, **Dynamic Input**, **Quick Properties**, and **Selection Cycling**. On starting AutoCAD Electrical, these tabs have default settings. You can change them according to your requirements. You will learn about them later in the book.

Snap and Grid

Grid lines are the checked lines on the screen at predefined spacing, see Figure-53. In AutoCAD Electrical, by default, the grids are displayed as checked lines. You can also display the grids as dotted lines, refer to Figure 4-39. To do so, set the 2D model space check box in the **Grid style** area. These dotted lines act as a graph that can be used as reference lines in a drawing. You can change the distance between grid lines as per your requirement. If grid lines are displayed within the drawing limits, it helps to define the working area. The grid also gives you an idea about the size of the drawing objects. To display the grid up to the limits, clear the **Display grid beyond Limits** check box. Now, the grids will be displayed only up to the limits set.

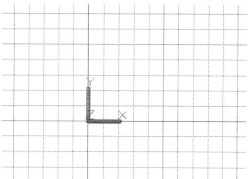
Figure-53. Grid

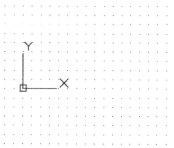
Figure-54. Dotted grid

Ortho Mode

If the **Ortho Mode** button is chosen, you can draw entities like lines at right angles only. You can use the F8 function key to turn ortho on or off.

Polar Tracking

- Click on the **Polar Tracking** button to restrict the cursor movements along the angle specified for polar tracking.
- To set the polar tracking angle, click on the down arrow next to the **Polar Tracking** button. A list of options will be displayed; refer to Figure-55.

Figure-55. Polar tracking angles

- Click on the desired angle option from the list. The cursor will start tracking as per the selected angle option.
- To specify the settings related to **Polar Tracking**, click on the **Tracking Settings** button from the list of options. The **Drafting Settings** dialog box will be displayed with the **Polar Tracking** tab selected; refer to Figure-56.

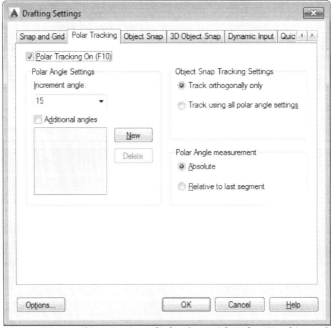

Figure-56. Drafting Settings dialog box with Polar Tracking tab

- Click in the **Increment angle** drop-down and select the desired angle.
- You can specify more than one angle for tracking. For that, click on the **Additional angles** check box and click on the **New** button. You will be asked to specify the additional angle.
- Enter the desired angle in the edit box. You can follow the same procedure to specify up to 10 additional angles.
- Click on the **OK** button to apply the specified settings.

Isometric Drafting

This button is used to enable isometric drafting. The procedure to use this option is given next.

- Click on the **Isometric Drafting** button from the Bottom bar. The orientation of the cursor will change as per the current isometric plane selected; refer to Figure-57.

Figure-57. Cursor orientation

- To select another isometric plane, click on the down arrow next to the **Isometric Drafting** button. A list of isometric planes will be displayed; refer to Figure-58.

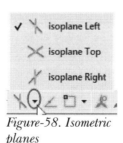

Figure-58. Isometric planes

- Select the desired isometric plane and the cursor will be oriented accordingly.

Annotation buttons

There is a group of four buttons that we call annotation buttons in AutoCAD Electrical; refer to Figure-59.

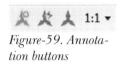

Figure-59. Annotation buttons

These buttons are used to manage annotation and their display in the drawing area. You will learn more about these options later in the book.

Workspace Switching

This flyout is used to switch between various workspaces available in AutoCAD Electrical. A workspace is defined as a customized arrangement of **Ribbon**, toolbars, menus, and window palettes in the AutoCAD Electrical environment. Click on the down arrow next to gear symbol in the Bottom bar. A flyout is displayed; refer to Figure-60.

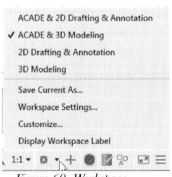

Figure-60. Workspace switching flyout

There are four default workspaces, **ACADE & 2D Drafting & Annotation**; **ACADE & 3D Modeling**; **2D Drafting & Annotation**; and **3D Modeling**. **ACADE & 2D Drafting & Annotation** workspace displays the tools and options related to 2D electrical drafting. **ACADE & 3D Modeling** workspace displays the tools and options related to basic 3D electrical modeling. In this environment, you can generate isometric drawings. The **2D Drafting & Annotation** workspace displays the tools to perform 2D drafting in general. The **3D Modeling** workspace displays almost all the tools required to perform advanced modeling.

Hardware Acceleration On

This button is used to set the performance of the software to an acceptable level.

Isolate Objects

This button is used to hide or isolate objects from the drawing area. On choosing this button, a flyout will be displayed with two options. Choose the **Isolate Objects** option from this flyout and then select the objects to hide or isolate. To end isolation or display a hidden object, click this button again and choose the **End Object Isolation** option.

Clean Screen

The **Clean Screen** button is at the lower right corner of the screen. This button, when chosen, displays an expanded view of the drawing area by hiding all the toolbars except the command window, Status Bar, and menu bar. The expanded view of the drawing area can also be displayed by choosing **View** > **Clean Screen** from the menu bar or by using the CTRL+0 keys. Choose the **Clean Screen** button again to restore the previous display state.

Customization

This option is used to customize the **Bottom** bar. You can add or remove the buttons from the **Bottom** bar as per the requirement. Click on the **Customization** button at the right corner in the **Bottom** bar. A flyout will display as shown in Figure-61. The options that are displayed with tick mark are available in the **Bottom** bar and other options are not. To display the other options or hide the current displaying options click on them in the flyout.

Figure-61. Customization flyout

For example, click on the **Units** option from the flyout. It will be added in the **Bottom** bar; refer to Figure-62. Now, you can change the unit style by clicking on this option in the Bottom bar any time while working.

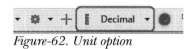
Figure-62. Unit option

Page left blank for Student Notes

Project Management

Chapter 3

The major topics covered in this chapter are:

- *Workflow in AutoCAD Electrical*
- *Starting a New Project*
- *Changing Properties of a project*
- *Adding drawings in the project*
- *Retagging and renumbering ladders in the drawings of project*
- *Plotting/publishing project files*

WORKFLOW IN AUTOCAD ELECTRICAL

AutoCAD Electrical is a software that can manage a complete project of electrical circuits. But, before we start to learn various tools in AutoCAD Electrical, it is necessary to understand the workflow in AutoCAD Electrical. The workflow in AutoCAD Electrical in given in Figure-1.

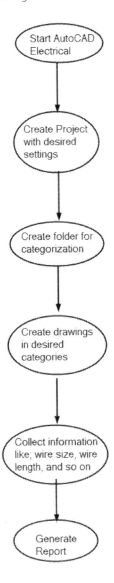

Figure-1. Workflow

We have learned about starting AutoCAD Electrical and now, its time to understand the procedures to manage projects in AutoCAD Electrical.

INITIALIZING PROJECT

As discussed earlier, Project is the categorized combination of various inter-related drawings. In AutoCAD Electrical, there are two major categories of drawings- Schematic diagrams and Panel drawings. Before we create these drawings, let's create a Project file. The procedure to create a project is given next.

- Start AutoCAD Electrical. If you are starting it for the first time then you will get the screen as shown in Figure-2. The marked left area displays the **Project Manager** which is used to manage projects.

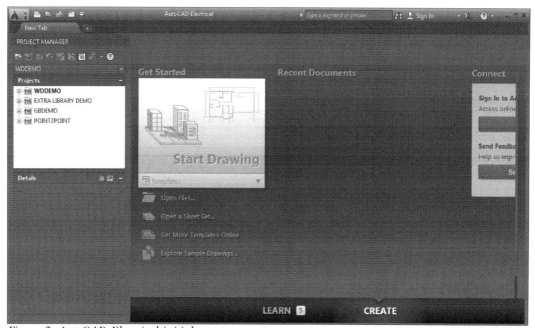

Figure-2. AutoCAD Electrical initial screen

- Click on the **Project Manager** button in the **Quick Access Toolbar**; refer to Figure-3, if the **Project Manager** is not displayed.

Figure-3. Project Manager button

- To be able to use all the tools in the **Project Manager**, open a drawing file or create a new one. The deactivated tools in the **Project Manager** will become active; refer to Figure-4.

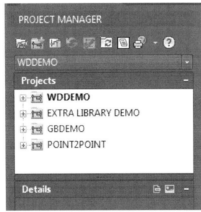
Figure-4. Project Manager

- Click on the **New Project** button in the **Project Manager**. The **Create New Project** dialog box will be displayed; refer to Figure-5.

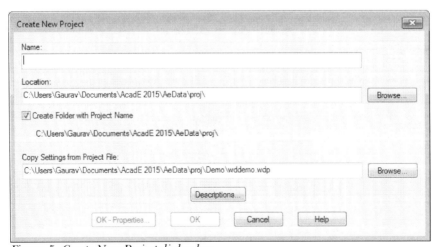
Figure-5. Create New Project dialog box

- Specify the name of the Project in the **Name** edit box of the dialog box.
- Make sure that the **Create Folder with Project Name** check box to create folders with the name of project file.
- Click on the **Browse** button next to **Location** edit box to change the location of the project files. The **Browse For Folder** dialog box will be display; refer to Figure-6.

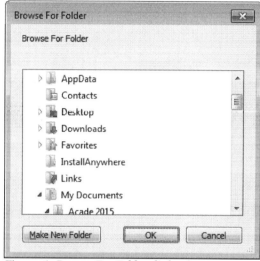

Figure-6. Browse For Folder dialog box

- Click on the nodes to move into sub-folder and select the desired location.
- Click on the **OK** button to create project files in the desired folder.
- If you want to copy settings from a desired project file then click on the **Browse** button next to **Copy Settings from Project File** edit box. Select the desired file using options in the dialog box displayed.
- Click on the **Descriptions** button to specify the description of the Project. The **Project Description** dialog box will be displayed as shown in Figure-7.

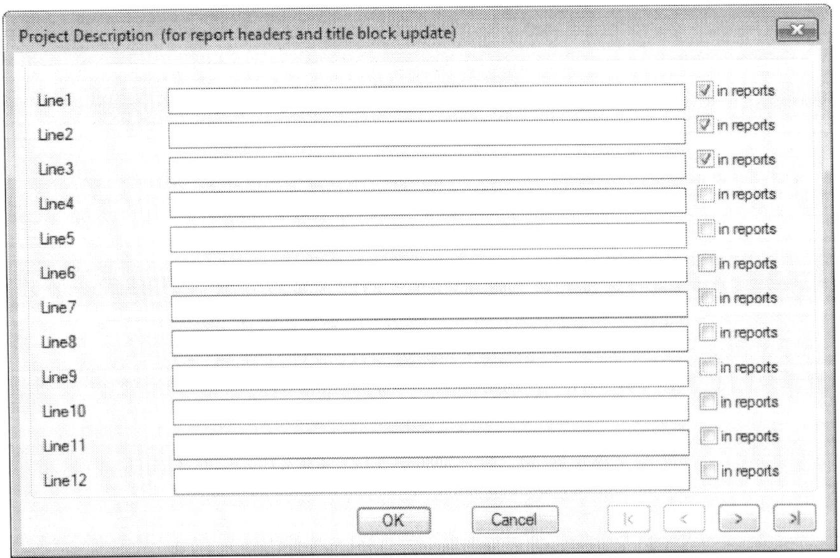

Figure-7. Project Description dialog box

- Specify the descriptions in the lines one by one. If you want to include any of the description line in your report then click on the **in reports** check box next that line.
- Click on the **OK** button to save the descriptions.
- Click on the **OK** button to create the Project. If you want to specify the properties of the project while creating, then click on the **OK - Properties** button. The **Project Properties** dialog box will be displayed; refer to Figure-8.

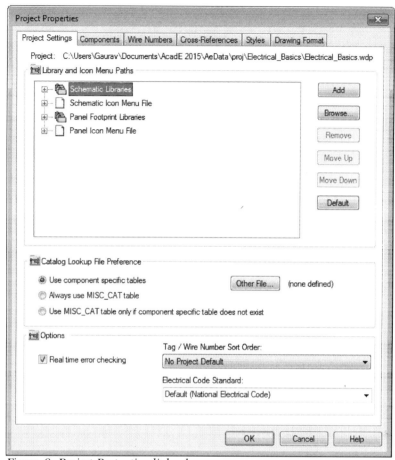

Figure-8. Project Properties dialog box

- Specify the desired settings and click on the **OK** button from the dialog box to apply settings. Note that you will learn about project properties in the next section.
- As you click on the **OK** button, the project name is added in the Project list in the **Project Manager**; refer to Figure-9.

Figure-9. Project name added in Project list

PROJECT PROPERTIES

Project Properties are the key players for any project whether it is for AutoCAD Electrical or any other related software. Follow the procedure given next to change the properties of the project.

- Right-click on the name of the project in the **Project Manager** whose properties are need to be changed. A shortcut menu will be displayed; refer to Figure-10.
- Click on the **Properties** button in shortcut menu. The **Project Properties** dialog box will be displayed as shown in Figure-8.

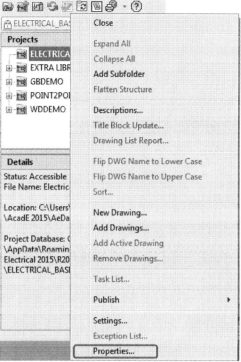
Figure-10. Shortcut menu

Library Icon Menu Paths area

- The first area in the dialog box is **Library Icon Menu Paths**. The options in this area are used to specify the directories containing symbols and icons required in Electrical drawings.
- To add any new library, click on the **Add** button from the dialog box. A new link will be added under the **Schematic Libraries** node. Specify the location of the directory that contains the symbols.
- Similarly, you can add more files for Icons and panels.

Catalog Lookup File Preference area

- Options in the **Catalog Lookup File Preference** area are used to set the database of files that are to be included in the library. If you add any new symbol or icon then make sure that you have entered it in the database file used in this section.

- To set a user defined database, click on the **Other File** button in this area. The **Catalog Lookup File** dialog box will be displayed as shown in Figure-11.

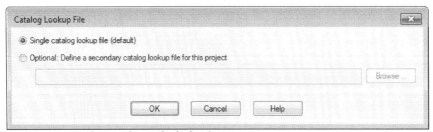

Figure-11. Catalog Lookup File dialog box

- By default system uses the default Lookup file available with the library. But if you need additional file for look up then click on the **Optional : Define a secondary catalog lookup file for this project** radio button and then click on the **Browse** button. The **Secondary Catalog Lookup File** dialog box will be displayed; refer to Figure-12.

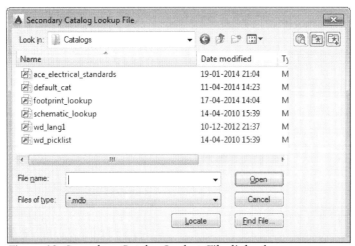

Figure-12. Secondary Catalog Lookup File dialog box

- Select the desired database file and click on the **Open** button. The location of file will be added in the edit box.
- Click on the **OK** button from the dialog box.

Options area

- Select the **Real time error checking** check box to allow the system to check for some basic errors in the drawing.

- The **Tag/Wire Number Sort Order** drop-down is used to specify the order in which the Tags/wire numbers will be arranged in the Project. There are many options in this drop-down to modify the order; refer to Figure-13.

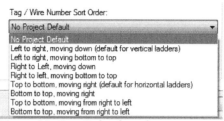

Figure-13. Sort Order drop-down

- **Electrical Code Standard** drop-down is used to set the standard for electrical codes.

OPENING A PROJECT FILE

The procedure to open a project file is similar to open a drawing file, which is given next.

- Click on the **Open Project** button from the **Project Manager**. The **Select Project File** dialog box will be displayed as shown in Figure-14.

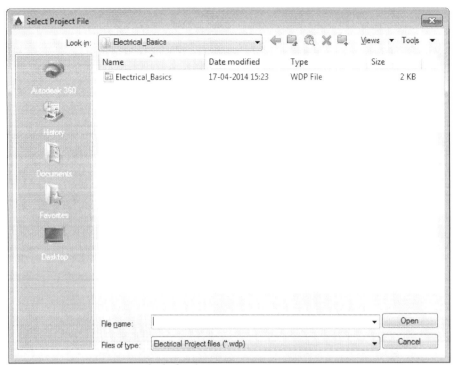

Figure-14. Select Project File dialog box

- Browse to the location of the project file and double-click on the file. The project file will open and display in the **Project Manager**.

NEW DRAWING IN A PROJECT

Creation of a new drawing is discussed earlier. At this stage, we will learn to create drawings in a project (Although earlier also, unknowingly we were creating drawing in a project). The procedure to create a drawing file in a project is given next.

- Click on the **New Drawing** button from the **Project Manager**. The **Create New Drawing** dialog box will be displayed as shown in Figure-15.

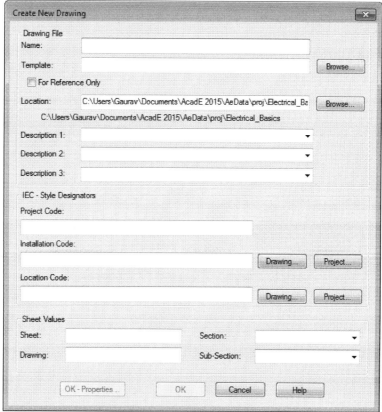
Figure-15. Create New Drawing dialog box

- Click in the Name edit box in the dialog box and specify the name of the drawing file.
- Click on the **Browse** button to select a file as template. The **Select template** dialog box will be displayed; refer to Figure-16.

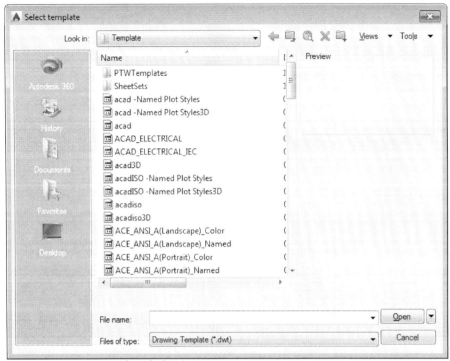

Figure-16. Select Template dialog box

- Select the desired template for your drawing. You can conclude the template style of default templates from their names in the list.
- After selecting the desired template, click on the Open button from the dialog box. Path of the template will be added in the **Template** edit box.
- Click in the **Description 1, Description 2,** and **Description 3** edit boxes one by one and specify the description as required.

The options in the **IEC - Style Designators** area are used to specify the identifiers for the project.

- Click in the **Project Code** edit box and specify the identifier code for the project.
- Similarly, you can specify **Installation Code** and **Location Code**. Note that you can use the codes of the earlier created drawings or projects by selecting the buttons corresponding to them.

- Also, you can specify the **Sheet, Drawing, Section** and **Sub-Section** numbers for the current drawing.
- All the parameters starting from **Project Code** to **Sub-Section** number are used as meta-data for the drawing and later help to identify the drawing.
- Click on the **OK** button. You are asked to set the default settings of the project or not. Click on **Yes** or **No** as desired. The drawing will be created.

REFRESH

The Refresh button is used to refresh the files in the Project. If you have performed any change in any drawing of the project then click on this button from the **Project Manager**, the information in the **Project Manager** will get updated.

PROJECT TASK LIST

The **Project Task List** button is used to display the tasks that are need to be performed in the current project. Click on this button, the list of tasks will be displayed.

PROJECT WIDE UPDATE OR RETAG

The **Project Wide Update/Retag** button is used to update or retag component tags, wire numbers, cross-references, and so on. The procedure to update or retag these parameters is given next.

- Click on the **Project Wide Update/Retag** button from the **Project Manager**. The **Project-Wide Update or Retag** dialog box will be displayed as shown in Figure-17.

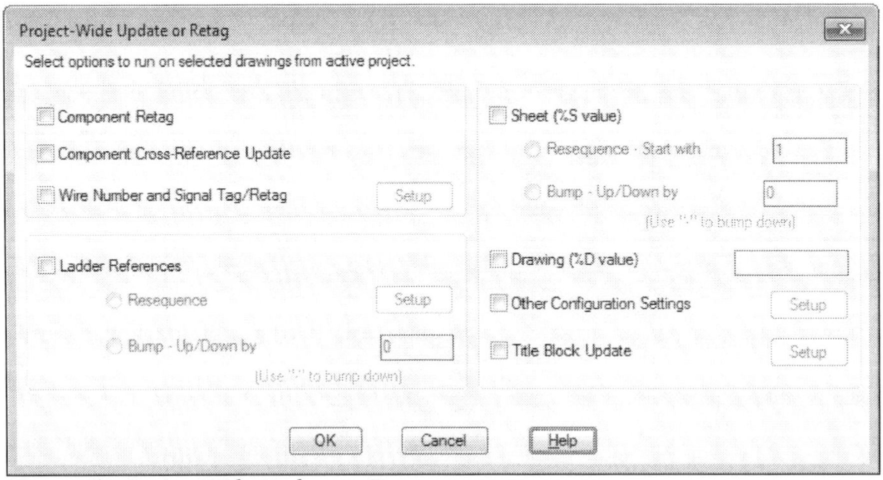

Figure-17. Project Wide Update or Retag

- Click on the **Component Retag** check box to specify tags of the non-fixed components while adding drawings in the project.
- Select the **Component Cross-Reference Update** check box to update all the component cross-references of the drawing while adding them in the project.
- Select the **Wire Number and Signal Tag/Retag** check box, the **Setup** button next to the selected check box will become active.

- Click on the **Setup** button to change the settings for wire numbers and signal tags. The **Wire Tagging (Project-Wide)** dialog box will be displayed; refer to Figure-18.

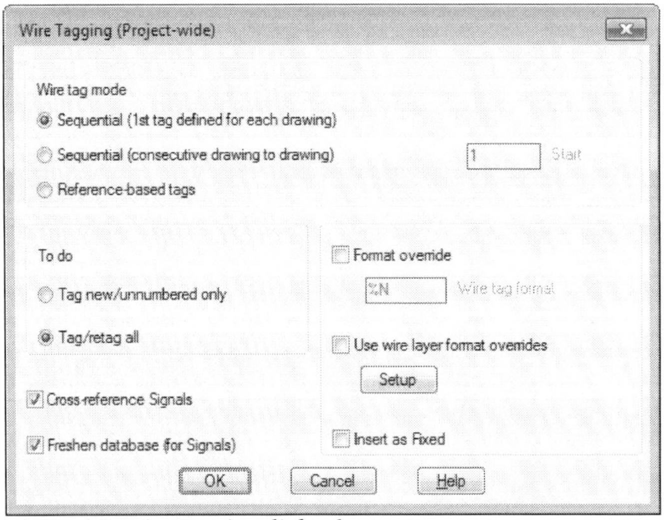

Figure-18. Wire Tagging dialog box

- Using the options in the dialog box you can update the tags of wires and signals. You will learn more about these options later in the book.

- Select the **Ladder References** check box to renumber the ladders in the circuits. On selecting this check box, the two radio buttons below it will become active.

- Select the **Resequence** radio button and click on the **Setup** button. The **Renumber Ladders** dialog box will be displayed.
- Specify the starting number of the ladders in the edit box displayed at the top in the dialog box. This numbering will be applicable to the first drawing in the project and similarly, you can specify the starting number of ladder for second onwards drawings by selecting the desired radio button in the dialog box.
- Click on the **OK** button from the dialog box to renumber the ladders in the current project.

- Select the **Bump - Up/Down by** radio button and specify the value in the edit box next to the radio button to increase the ladder numbers by specified value. Specify the negative value to decrease the ladder numbers.
- Select the desired check box from the right area of the dialog box and specify the related values.
- Click on the **OK** button from the dialog box. The **Select Drawings to Process** dialog box will be displayed refer to Figure-19.

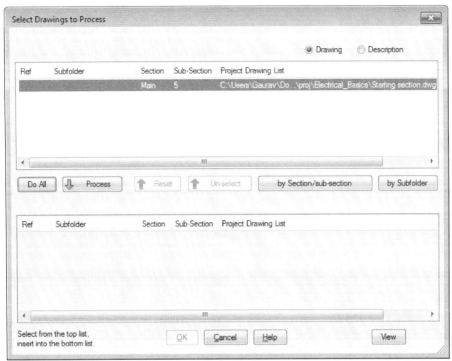

Figure-19. Select Drawings to Process dialog box

- Select the drawings from the list by holding the CTRL key and then click on the **Process** button to include the drawings for applying the settings we have specified earlier. The drawings will come in the bottom list.
- You can include all the drawings by clicking on the **Do All** button from the dialog box.
- Click on the **OK** button from the dialog box, the modified settings will be applied to the selected drawings.

DRAWING LIST DISPLAY CONFIGURATION

This button is used to configure the details of the drawings that are to be displayed in the **Project Manager** with the name of the drawings. The procedure to use this option is given next.

- Click on the **Drawing List Display Configuration** button. The **Drawing List Display Configuration** dialog box will be displayed as shown in Figure-20.

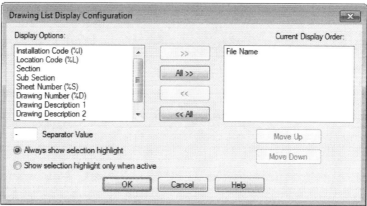

Figure-20. Drawing List Display Configuration

- Select the details from the left list box that you want to display in **Project Manager** one by one.
- Press the **Include** button from the dialog box. The details will be added in the right list box.
- You can include all the details by clicking on the **All>>** button.
- To exclude any detail or all the details, click on the **<<** or **<<All** button respectively.
- After selecting the desired details, click on the **OK** button from the dialog box. The details will be displayed with the name of the drawing in the **Project Manager**.

PLOTTING AND PUBLISHING

After creating any drawing in the software, the next step is to take a hardcopy or distribution media by which the drawing can be shared with other concerned people. There are four tools to plot or publish drawings of a project: **Plot Project, Publish to WEB, Publish to PDF/DWF/DWFx,** and **ZIP Project**; refer to Figure-21. These tools are one by one discussed next.

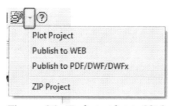

Figure-21. Tools to plot publish

Plot Project

The **Plot Project** tool is used to plot drawings of the project on the paper by using your printing/plotting device. The steps to use this tool are given next.

- Click on the **Plot Project** tool from the **Project Manager**. The **Select Drawings to Process** dialog box will be displayed; refer to Figure-19.
- Include the drawings that you want to plot by using the **Process** button after selecting them in the list.
- Click on the **OK** button from the **Select Drawings to Process** dialog box. The **Batch Plotting Options and Order** dialog box will be displayed; refer to Figure-22.

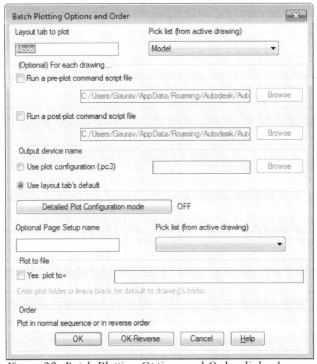

Figure-22. Batch Plotting Options and Order dialog box

- Click on the **Detailed Plot Configuration mode** button to specify settings for the plot. The **Detailed Plot Configuration Option** dialog box will be displayed; refer to Figure-23.

Figure-23. Detailed Plot Configuration Option dialog box

- Modify the desired settings of plotting like orientation of page, plot area, paper size and so on using the options in this dialog box.
- To select the desired paper size, click on the **Pick from generic list** radio button in the **Papersize** area. The dialog box with various paper sizes will be displayed; refer to Figure-24.
- Select the desired paper size and click on the **OK** button from the dialog box.
- Next, click on the **On** button from the **Detailed Plot Configuration Option** dialog box.

Figure-24. Paper sizes

- Click on the **OK** button from the **Batch Plotting Options and Order** dialog box. The drawings of the project will be plotted by the assigned plotter.

Publish to WEB

This tool is used to publish drawings on the website. The procedure to use this tool is given next.

- Click on the **Publish to WEB** tool from the **Plot/Publish** drop-down in the **Project Manager**. The **Publish to WEB** dialog box will be displayed as shown in Figure-25.

Figure-25. Publish to WEB dialog box

- The edit box in this dialog box will be used to specify the storage location of web files of the drawing. Click on the Browse button next to the edit box. The **Browse For Folder** dialog box will be displayed; refer to Figure-26.

Figure-26. Browse For Folder dialog box

- Browse to the desired folder and click on the **OK** button from the dialog box to exit.
- Click in the **Image** drop-down and select the desired image format from **DWF, JPEG,** and **PNG** for web view.
- Click on the **OK** button from the dialog box. The **Select Drawings to Process** dialog box will be displayed as discussed earlier.
- Include the desired drawings and click on the **OK** button from the dialog box. The **AutoCAD Electrical Publish to Web** dialog box will be displayed; refer to Figure-27.

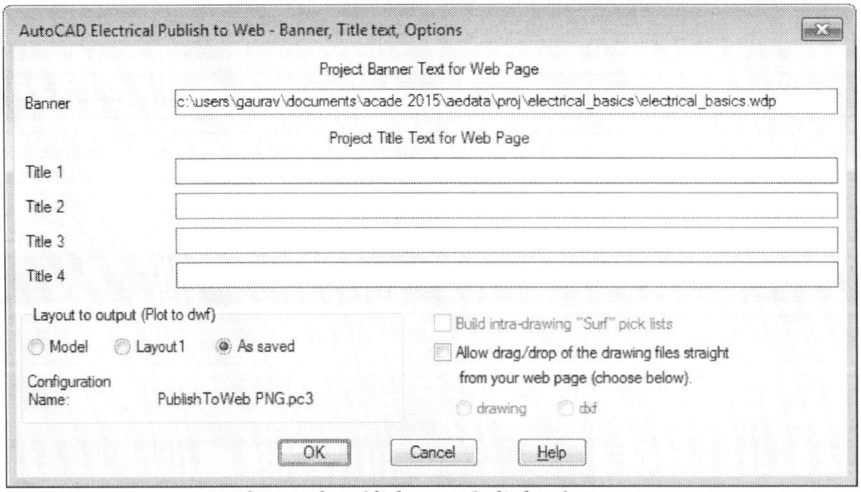

Figure-27. AutoCAD Electrical Publish to Web dialog box

- Specify the meta data as desired in the displayed fields.
- Select the **Allow drag/drop of the drawing files straight from your web page** check box to enable drag/drop of the drawing files.
- Click on the **OK** button from the dialog box. After the publishing is finished, the **Finish** page of the **Publish to Web** dialog box will be displayed as shown in Figure-28.

Figure-28. Publish to WEB dialog box Finish page

- Click on the **Preview** button to preview the web page of the drawing.
- Click on the **OK** button from the dialog box to exit.

Publish to DWF/PDF/DWFx

This tool is used to publish the drawings in Dwf or pdf forms. The procedure to use this tool is given next.

- Click on the **Publish to DWF/PDF/DWFx** tool from the **Plot/Publish** drop-down in the **Project Manager**. The **Select Drawings to Process** dialog box will be displayed as discussed earlier.
- Include the drawings that you want to publish in the DWF or PDF format. Click on the **OK** button from the dialog box. The **AutoCAD Electrical-Publish Setup** dialog box will be displayed as shown in Figure-29.

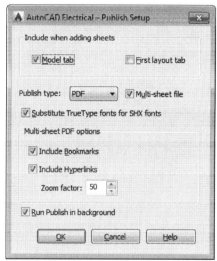

Figure-29. AutoCAD Electrical Publish Setup dialog box

- Select the check boxes for the options that you want to include in your DWFs and PDFs.
- Click in the **Publish type** drop-down to change the type of file for publishing.
- Click on the **OK** button from the dialog box. The **Publish** dialog box will be displayed; refer to Figure-30.

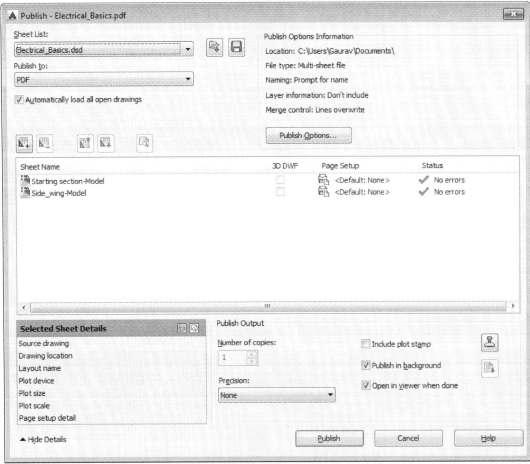

Figure-30. Publish dialog box

- Set the desired options and then click on the **Publish** button from the dialog box. The **Specify PDF File** dialog box will be displayed (in case of PDF format selected); refer to Figure-31.

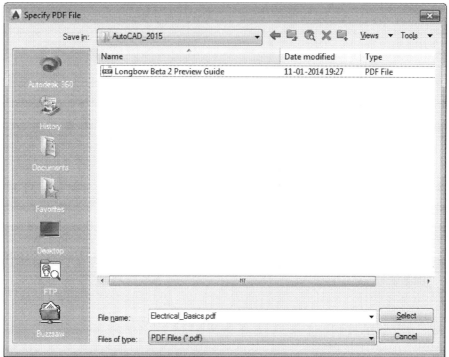
Figure-31. Specify PDF File dialog box

- Browse to the desired location and then click on the **Select** button. If you have unsaved sheet list then the **Publish - Save Sheet List** dialog box will be displayed as shown in Figure-32.

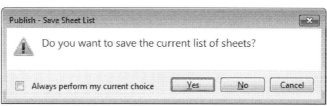
Figure-32. Publish Save Sheet List dialog box

- Click on the **Yes** button from the dialog box. The dialog box to specify location for saving the file will be displayed. Browse to the desired location and save the file.
- The drawing file will be saved in the PDF format and will automatically open in the PDF reader of your system.

Zip Project

This tool works in the same way as the other publishing tools discussed above. The procedure to use this tool is given next.

- Click on the **Zip Project** tool from the **Plot/Publish** drop-down in the **Project Manager**. If you are creating the zip file for the first time in this version of AutoCAD, then you will get the error message as shown in Figure-33.

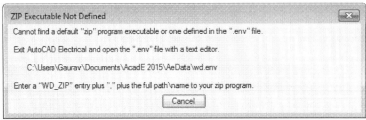

Figure-33. Error message

- Click on the **Cancel** button from this message box and exit AutoCAD Electrical.
- Open the folder **C:\Users\{your user name}\Documents\AcadE2015\AeData** using windows browser.
- Open the **wd.env** file using any text editor(in our case, Notepad). The file will be displayed as shown in Figure-34.

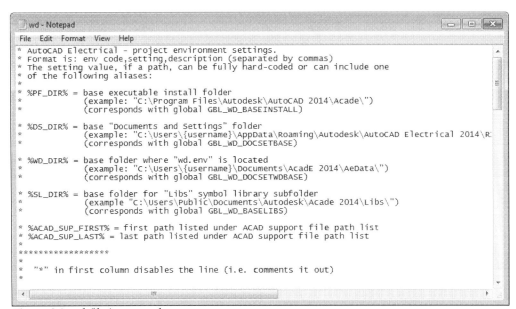

Figure-34. wd file in notepad

- Move down in the file at the location as shown in Figure-35.

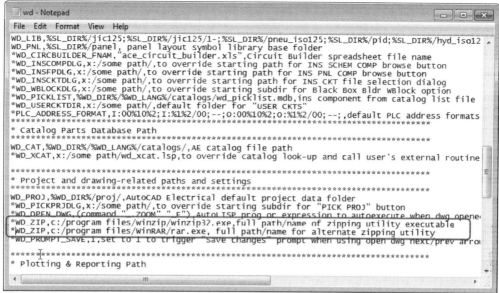

Figure-35. Location to identified

- Check carefully the lines marked in the red box. You will notice that these lines link the zip file with two programs **WinZip32** and **WinRAR**.
- Remove the ***** mark from these line. Press **CTRL+S** to save the file.
- Restart the AutoCAD Electrical and reopen the project file and drawing you were working on.
- Click on the **Zip Project** tool from the **Plot/Publish** drop-down in the **Project Manager**. The **Select Drawings to Process** dialog box will be displayed as discussed earlier.
- Rest of the procedure is same as discussed earlier.

REMOVING, REPLACING, AND RENAMING DRAWINGS IN A PROJECT

Till this point you have learned to add drawings in the Project and you have learned to modify properties of the drawings. Now, you will learn to remove, replace, or rename a drawing file in a project. The steps to perform these actions are given next.

- Select the drawing from the **Project Manager** that you want to remove, replace, or rename. Right-click on it, a shortcut menu will be displayed; refer to Figure-36.

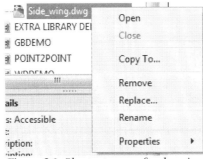

Figure-36. Shortcut menu for drawing

- Select the **Remove, Replace,** or **Rename** option from the menu.
- Follow the instructions and do accordingly.

- If you have selected the **Remove** option then a confirmation will be asked and if you select Yes then the file will be removed.

- If you have selected the **Replace** option then the dialog box will be displayed prompting you to select the drawing for replacement.

- If you have selected the **Rename** option then you will be prompted to specify the new name of the drawing.

Inserting Components

Chapter 4

Topics Covered

The major topics covered in this chapter are:

- **Inserting Components using Icon menu**
- **Inserting Components using Catalog Browser**
- **Inserting Components using User Defined list**
- **Inserting Components using Equipment list**
- **Inserting Components using Panel list**
- **Inserting Components using Terminal (Panel list)**
- **Pneumatic, Hydraulic, and P&ID components**

ELECTRICAL COMPONENTS

Electrical Components in AutoCAD Electrical are schematic symbols of various components used in electrical systems. There are various ways to insert the electrical components in AutoCAD Electrical. Some of the methods to insert electrical components are given next.

INSERTING COMPONENT USING ICON MENU

This tool is used to insert schematic symbols of various electrical components by using the **Icon Menu**. The procedure to use this tool is given next.

- Click on the Down arrow below **Icon Menu** button in the **Schematic** tab of the **Ribbon**. The list of tools will be displayed; refer to Figure-1.

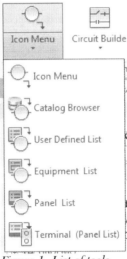

Figure-1. List of tools

- Click on the **Icon Menu** tool in this menu. The **Insert Component** window will be displayed; refer to Figure-2.

AutoCAD Electrical 2015 Black Book   4-3

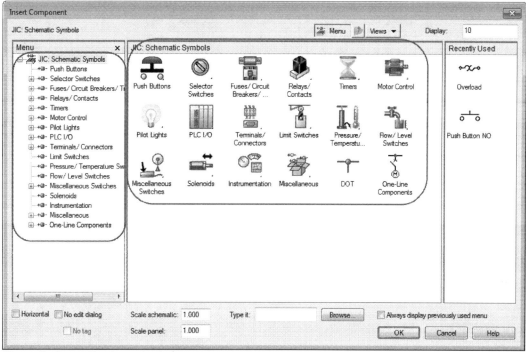

Figure-2. Insert Component window

- Select the category of desired component from the Menu box at the left in the dialog box. Note that the categories that have a **+** sign before their name are having sub-categories to browse in.
- After selecting the category and subcategory, click on the desired symbol. The symbol will be attached to the cursor and you will be prompted to specify the location for the symbol; refer to Figure-3.

Figure-3. Component attached to cursor

- Click in the drawing area to place the component symbol at desired place. The **Insert/Edit Component** dialog box will be displayed; refer to Figure-4.

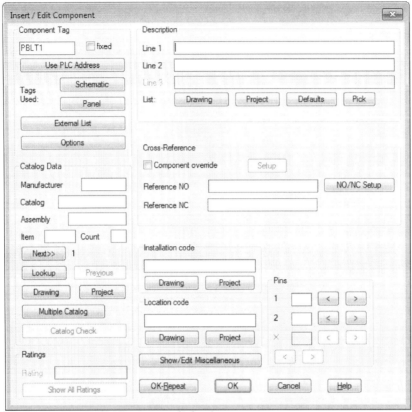

Figure-4. Insert Edit Component dialog box

- This dialog box is divided into various areas like; **Component Tag**, **Description**, **Catalog Data**, **Cross-Reference** and so on. The options in these areas specify various parameters of the component being inserted. We will start with the **Component Tag** area and then one by one we will use options in other areas.

Component Tag area

- The edit box in this area is used to specify the tag value for your current component. Click in the edit box and specify the desired value.
- You can link your component with PLC by using the tag of PLC. To do so, click on the **Use PLC Address** button. If

your components are not directly connected to PLC via wires as in our case then the **Connected PLC Address Not Found** dialog box will be displayed; refer to Figure-5.

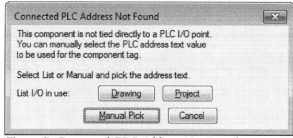

Figure-5. Connected PLC Address Not Found

- Click on the **Drawing** button to display all the PLC connections available. The **PLC I/O Point List (this drawing)** dialog box will be displayed as shown in Figure-6.

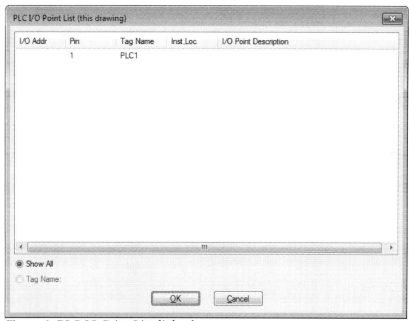

Figure-6. PLC IO Point List dialog box

- Select the desired plc pin from the dialog box and click on the **OK** button from the dialog box. The component will be linked to the selected pin of plc.

Schematic tag

- Using the **Schematic** button in the **Component Tag** area. The **SOL Tags in Use** dialog box will be displayed as shown in Figure-7.

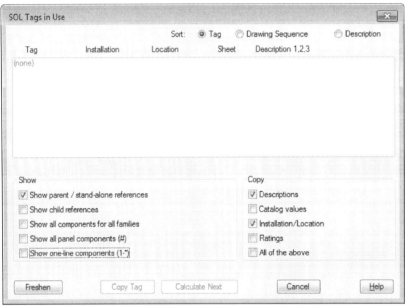

Figure-7. SOL Tags in Use dialog box

- Select the **Show all components for all families** check box. The list of all the tags in the drawing will be displayed. Also the name of the dialog box will be modified as shown in **All Tags in Use**.
- Select the desired tag number and then select **Copy Tag** or **Calculate Next** button. If you select the **Copy Tag** button then same tag will be assigned to the current component. If you select the **Calculate Next** button then the next tag number will be assigned to the component. Note that in both the cases system will ask your permission to overwrite the tag value. Do as per your requirement.

In the same way, you can use the **Panel** button and the **External List** button from the **Component Tag** area.

Catalog Data area

The options in this area are used to link the inserted component to a catalog. A catalog is a collection of various related components in a categorized form. To specify the details of the Catalog Data follow the steps given next.

- Click in the **Manufacturer** edit box of the **Catalog Data** area and specify the name of the manufacturer. There are various companies like **AB, ABB, Siemens,** and so on that manufacture electrical components are available in the AutoCAD Electrical Database. You can also start a new manufacturer by specifying your desired name.
- Click in the **Catalog** edit box and specify the value of Catalog number. This number can include alphabets as well as numeric. For example, 300F-1PB.
- Click in the **Assembly** edit box and specify the desired value for assembly code.
- Click in the **Item** edit box and specify the item number for the component. Note that each type of component has a unique Item number.
- Click in the **Count** edit box and specify the number of components required in the current drawing. The specified numbers will be automatically added in the Bill of Materials. The area after specifying all the values will be displayed as shown in Figure-8.

Figure-8. Catalog Data area

- Instead of specifying all the parameters related to Catalog data, you can pick the desired component's details from the library of standard components in AutoCAD Electrical. To do so, click on the **Lookup** button in this area. The **Catalog Browser** dialog box will be displayed as shown in Figure-9.

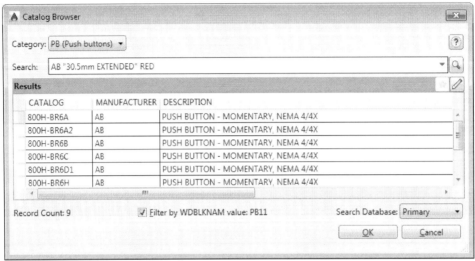

Figure-9. Catalog Browser dialog box

- Select the desired option from the **Category** drop-down at the top in the dialog box. If you want to check all the components in this category then click on the search button. The list of all the components in that category will be displayed.
- You can customize user search by specifying the desired keywords in the **Search** edit box. After specifying the keywords, click on the **Search** button. The list of components related to the specified keyword will be displayed.
- Double click on the desired component in the list. The related data will be displayed in the **Catalog** area.
- Similarly, you can use **Drawing** button or **Project** button to select the catalog data from earlier created components in the drawing or project respectively.

Multiple Catalog Data

- You can use multiple catalog for a component to display alternates for the component. To do so, click on the **Multiple Catalog** button from the **Insert/Edit Component** dialog box. The **Multiple Bill of Material Information** dialog box will be displayed; refer to Figure-10.

Figure-10. Multiple Bill of Material Information dialog box

- Enter the details of **Manufacturer**, **Catalog**, and **Assembly** in the related edit boxes. Now, you have specified first manufacturer of the desired component.
- Click on the **Sequential Code** drop-down and select **02** from the list displayed; refer to Figure-11.

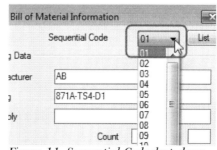

Figure-11. Sequential Code drop down

- Specify the alternate data for **Manufacturer, Catalog,** and **Assembly**. Repeat the process until you get the desired number of alternates for the selected type of components. You can specify 99 alternates at max.
- After specifying the data, click on the **OK** button from the dialog box.

Description Area

The options in this area are used to specify the description about the component. To specify the description, follow the steps given next.

- Click in the **Line 1** edit box in this area and specify the first line of description. Similarly, click in the **Line 2** edit box and specify the second line of description.
- If you want to use description of any component in current drawing or project then click on the **Drawing** button or **Project** button respectively.
- The related dialog boxes will be displayed and you will be prompted to select the description of a component from the list. Select the desired component description and click on the **OK** button from the dialog box.
- You can also select the description from the default description list. To do so, click on the **Defaults** button in this area. The **Descriptions** dialog box will be displayed; refer to Figure-12.

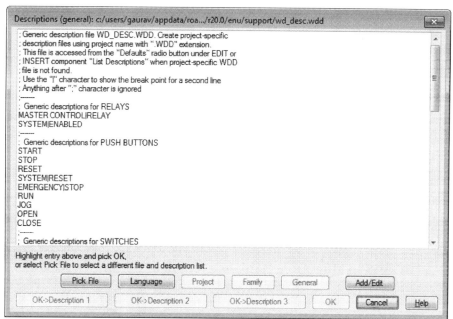

Figure-12. Descriptions dialog box

- Double click on any of the description in this dialog box. Note that the lines with all capital characters are counted in the description. So, you need to double click only on those lines.

Cross-Reference Area

This is a very important area of the dialog box. Using the options in this area, you can link the pins of your component with the other components the drawing/project. The procedure to use options in this area is given next.

- When we insert a component in AutoCAD Electrical, then based on the already existing components, the inserted component is automatically referenced to existing components. This reference is of a certain reference format. To override this format, click on the **Component override** check box. The **Setup** button next to it will get activated.
- Click on the **Setup** button. The **Cross-Reference Component Override** dialog box will be displayed; refer to Figure-13.

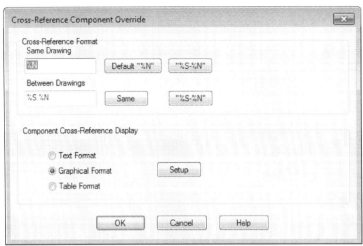

Figure-13. Cross Reference Component Override dialog box

- Using the options in this dialog box, you can change the reference style. By default, for same drawing %N is used as reference number and for Between drawings %S.%N is used. Glossary is given next.

%S Sheet number of the drawing
%D Drawing number
%N Sequential or reference-based number applied to the component
%X Suffix character position for reference-based tagging (not present = end of tag)
%P IEC-style project code (default for drawing)
%I IEC-style "installation" code (default for drawing)
%L IEC-style "location" code (default for drawing)

- Specify the desired identifiers.
- You can also change the display format by using the radio buttons in the **Component Cross-Reference Display** area of the dialog box.
- After specifying the desired format style, click on the **OK** button from the dialog box to exit.

- Each component need to be specified by NC or NO contact points. NC means Normally Closed and NO means Normally Open. Using the Reference NO and Reference NC edit boxes, you can specify the references for the contact types.

- A switch generally has two pins, but if you are working on PLC or other integrated circuit components then you need to setup NC/NO for all the pins of components. To do so, click on the **NO/NC Setup** button. The **Maximum NO/NC counts and/or allowed Pin numbers** dialog box will be displayed; refer to Figure-14.

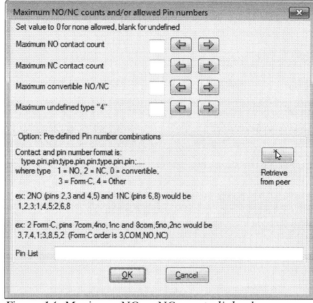

Figure-14. Maximum NO or NC counts dialog box

- Specify the numbers of contact types for different categories in this dialog box.
- In the **Pin List** edit box, you can define the connectivity of one pin with the other for NC, NO, COM, or others. Check the description in **Option** area of the dialog box to know the method of linking pins for switches.

Installation Code and Location Code

- The edit boxes in these areas are used to specify the installation code and location code for the current component. The installation code is used to specify the line or circuit identifier in which you want to add the component. For example, you can specify **Line1 230VAC** as installation code.
- **Location code** is used to specify the identifier for physical location of the component. For example, you can specify **Sub-Panel 20 Pin** as location code for the component.

Pins area

The options in this area are used to number the pins of the component so that later they can be connected with other components in the line. The procedure to use these options is given next.

- Click on the **<** button or the **>** button to decrease or increase the pin number.

- Click on the **OK** button from the dialog box to exit. The component will be displayed with the specified parameters in the drawing.

CATALOG BROWSER

The **Catalog Browser** is the standard library of various components used in electrical engineering. You can insert any desired component using the options in the **Catalog Browser**. The procedure to use the **Catalog Browser** is given next.

- Click on the down arrow below **Icon Menu** button in the **Ribbon,** the Component drop-down will be displayed. Click on the **Catalog Browser** tool from the list; refer to Figure-15. The **Catalog Browser** will be displayed; refer to Figure-16.

Figure-15. Catalog Browser button

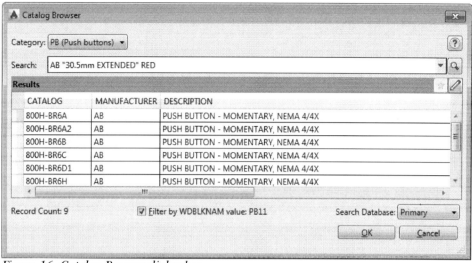

Figure-16. Catalog Browser dialog box

- Click on the edit box for Category, a list of various component categories will be displayed; refer to Figure-17.

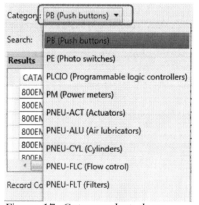

Figure-17. Category drop-down

- Select the desire component type from the list.
- Click in the Search edit box and specify the requirements in keywords separated by space.
- Click on the **Search** button next to **Search** edit box, the list of components matching the keywords will be displayed; refer to Figure-18.

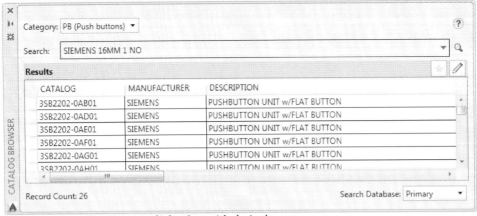

Figure-18. Catalog Browser dialog box with desired components

- Click on the component that you want to include in your drawing, a small toolbar will be displayed; refer to Figure-19.

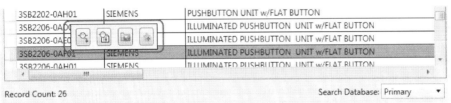

Figure-19. Toolbar in Catalog Browser

- The [icon] is used to specify a symbol for the current selected component. The [icon] is used to display the assembly details of the selected component. The [icon] is used to display the details of the current selected component in the web browser. Using this button, you will reach to the manufacturer's website. The [icon] button is used to add the current component in the favorite components list.
- Click on the [icon] button to assign a symbol. The **Insert Component** dialog box will be displayed; refer to Figure-20.

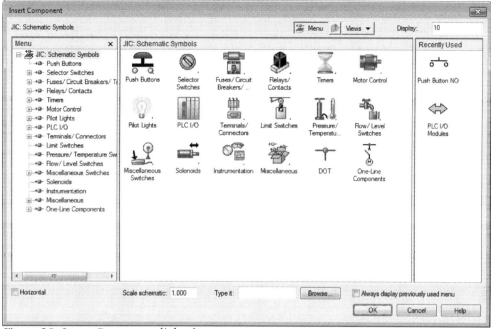

Figure-20. Insert Component dialog box

- Click on the category for which your component resembles. In our case, Push Buttons. The list of symbols will be displayed.
- Select the desired button from the list of symbols, in our case, **Illuminated Push Button NO**. The **Assign Symbol To Catalog Number** dialog box will be displayed (if the catalog number is unassigned); refer to Figure-21.

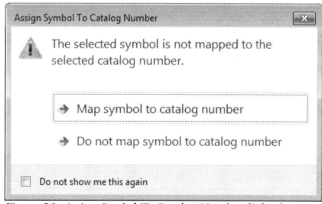

Figure-21. Assign Symbol To Catalog Number dialog box

- Click on the **Map symbol to catalog number** button from the dialog box to map the symbol for the catalog. The symbol will get attached to the cursor.
- Click on the electrical line to which you want to connect the component or click in the drawing area to place the component; refer to Figure-22. Note that after placing the component, the **Insert/Edit Component** dialog box will be displayed as discussed earlier.

Figure-22. Component after placing on line

- Specify the desired parameters and click on the **OK** button from the dialog box.
- If you again go to **Catalog Browser** and click on the same component, then you will get symbols mapped to the component; refer to Figure-23.

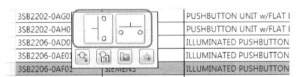

Figure-23. Symbols mapped to component

USER DEFINED LIST

The **User Defined List** tool is used to display the list of components that are collected by user under a common list. The procedure to use this tool is given next.

- Select the **User Defined List** tool from the Component drop-down. The **Schematic Component or Circuit** dialog box will be displayed as shown in Figure-24.

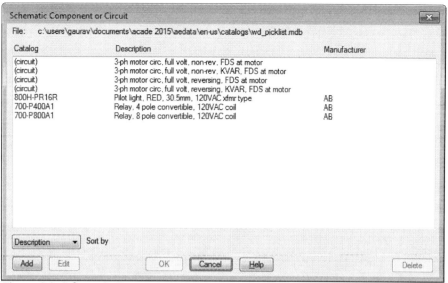
Figure-24. Schematic Component or Circuit dialog box

- Select any component/circuit from the list and click on the **OK** button from the dialog box. The component/circuit will get attached to the cursor. Click to place the component/circuit.
- You can add the desired components/circuits in the list by using the **Add** button. Click on the **Add** button. On doing so, the **Add record** dialog box will be displayed; refer to .

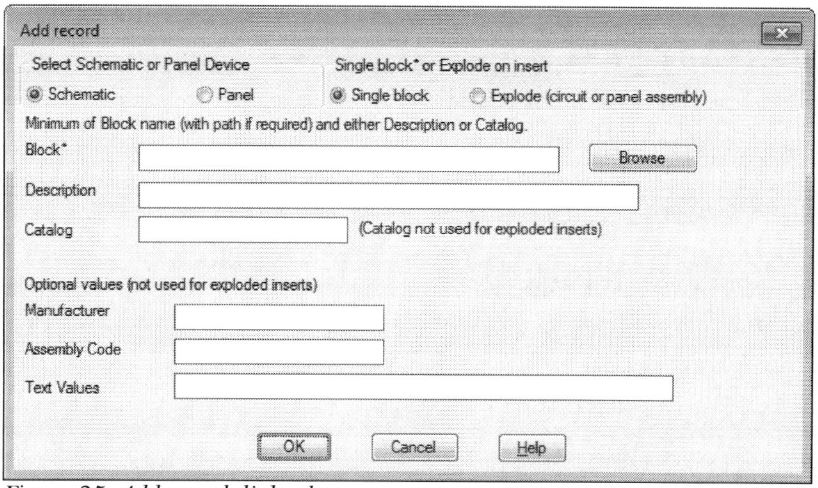

Figure-25. Add record dialog box

- Select the **Schematic** or **Panel** radio button to specify the type of component we are going to add in the library.
- If it is a single block then select the **Single block**

radio button otherwise select the **Explode (Circuit or panel assembly)** radio button.
- Click on the **Browse** button to specify the location of block for component. The **Select Schematic component or circuit** dialog box will be displayed as shown in Figure-26.

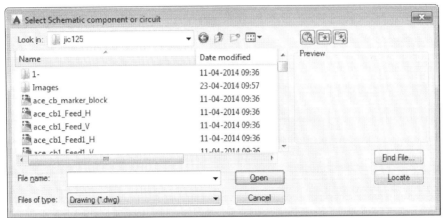

Figure-26. Select Schematic component or circuit dialog box

- Select the drawing file of the symbol that you want to assign for your component. Click on the **Open** button from the dialog box. Note that you can make your own symbol by using the drawing tools and after saving it at desired location, you can add it to the library.
- Click in the **Description** edit box and specify the description from the component.
- Click in the **Catalog** edit box and specify the catalog identifier for the component.
- Click in the **Manufacturer** edit box and specify the name of the manufacturer.
- Click in the **Assembly Code** edit box and specify the desired assembly code for the component.
- Click in the **Text Values** edit box to specify the text identifiers for the component.
- Click on the **OK** button from the dialog box to place the newly added symbol.

EQUIPMENT LIST

The **Equipment List** tool is used to display or add the components in the drawing. The procedure to use this tool is given next.

- Click on the **Equipment List** tool from the **Component** drop-down. The **Select Equipment List Spreadsheet File** dialog box will be displayed; refer to Figure-27.

Figure-27. Select Equipment List Spreadsheet File dialog box

- Select the desired database file using the options in the dialog box. Click on the **Open** button from the dialog box. The **Table Edit** dialog box will be displayed; refer to Figure-28.

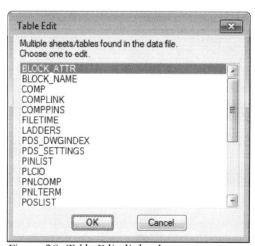

Figure-28. Table Edit dialog box

- Select the desired data table from the list and click on the **OK** button from the dialog box. The **Settings** dialog box will be displayed; refer to Figure-29.

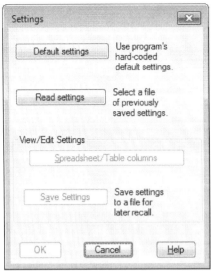
Figure-29. Settings dialog box

- Click on the **Default settings** button to use the default settings of the program. You can use the settings of an already existing **.wde** file by using the **Read settings** button. *.wde extension file is used to specify Equipment list setup file.
- After specifying the settings file, the **Spreadsheet/Table columns** button will become active in the **View/Edit Settings** area of the dialog box. Click on the **Spreadsheet/Table columns** button. The **Equipment List Spread Settings** dialog box will be displayed; refer to Figure-30.

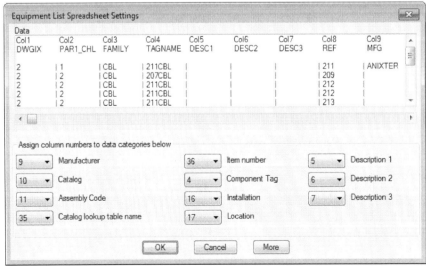
Figure-30. Equipment List Spreadsheet Settings dialog box

- Change the data category of the columns as per the requirement by using the drop-downs in this dialog box. After modifying the category, click on the **OK** button from the dialog box. Click on the **OK** button from the next dialog box. The **Schematic equipment in** dialog box will be displayed; refer to Figure-31.

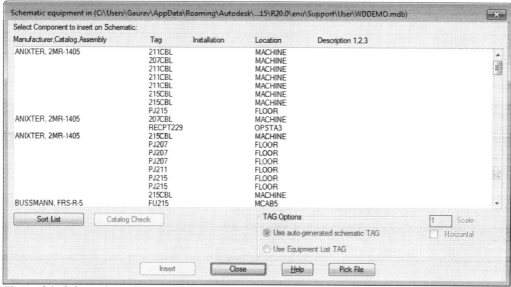
Figure-31. Schematic equipment in dialog box

- Browse through the list of equipment and click on the desired equipment from the list.
- Click on the **Insert** button from the dialog box. If there is no symbol assigned to the component then the **Insert** dialog box will be displayed as shown in Figure-32.

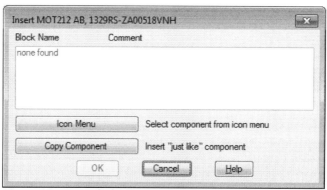
Figure-32. Insert dialog box

- Click on the Icon Menu button and assign the desired

symbol to the component. The symbol will get attached to the cursor.
- Click in the drawing area or on the electrical line to place the component/equipment. The component/equipment will be placed and the **Insert/Edit Component** dialog box will be displayed as discussed earlier.
- Specify the desired parameters and click on the **OK** button from the dialog box to exit.

PANEL LIST

The Panel List tool is used to insert the parts from panels. The procedure to use this tool is given next.

- Click on the **Panel List** tool from the **Component** drop-down. The **Panel Layout List** dialog box will be displayed as shown in Figure-33.

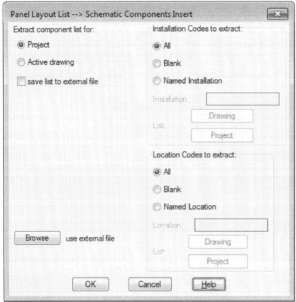

Figure-33. Panel Layout List dialog box

- You can extract the list of components from the current project or active drawing. Select the **Project** radio button or **Active drawing** radio button to get the list of components.

- If you want to save the extracted list of components in an external file, select the **save list to external file** check box.
- You can also use an external file for equipment list by using the **Browse** button.
- After specifying the desired parameters, click on the **OK** button from the dialog box. If you have selected the **Project** radio button then the **Select Drawings to Process** dialog box will be displayed as discussed earlier.
- Include the drawings from which you want to extract the list of components. Click on the **OK** button from the dialog box. The list of components will display.

Note that the **Panel List** and **Terminal (Panel List)** tools work only if we have panels in our current project or drawing. We will learn more about these tools later in the book.

There are three more categories of components that are available in AutoCAD Electrical; Pneumatic Components, Hydraulic Components, and P&ID Components. The use of these components and their procedure of insertion in drawing is given next.

PNEUMATIC COMPONENTS

Pneumatic components are those components that work because of pressure of the gases flowing through them. In some electrical systems, you might need to deal with pneumatic components also. So, AutoCAD Electrical provides a special database of these components. The procedure to insert the Pneumatic components is given next.

- Click on the Down arrow next to **Insert Component** panel name; refer to Figure-34. The expanded **Insert Component** panel will be displayed; refer to Figure-35.

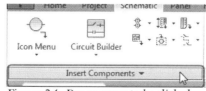

Figure-34. Down arror to be clicked

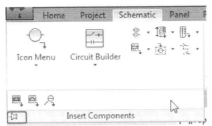

Figure-35. Expanded Insert Components panel

- Click on the pin button (in red box in the above figure) to keep the panel expanded.
- Click on the **Insert Pneumatic Components** button. The **Insert Component** dialog box will be displayed as shown in Figure-36.

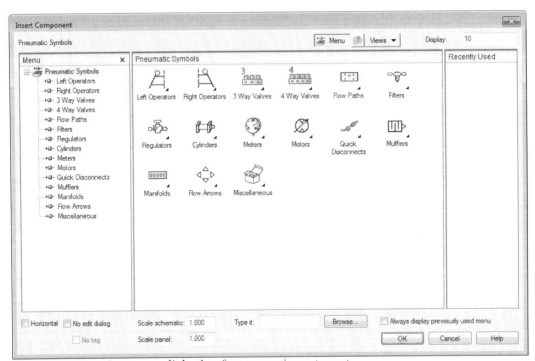

Figure-36. Insert Component dialog box for pneumatic components

- Click on the desired category and select the component that you want to insert in the drawing. The component will get attached to the cursor.
- Click in the drawing area to place the component. The **Insert/Edit Component** dialog box will be displayed.

- Specify the parameters as discussed earlier and click on the **OK** button from the dialog box. The component will be placed with the specified parameters.

HYDRAULIC COMPONENTS

The hydraulic components are those components that work due to the pressure of fluids flowing through them. The procedure to insert the hydraulic components is given next.

- Click on the **Insert Hydraulic Components** tool from the expanded **Insert Components** panel. The **Insert Component** dialog box with hydraulic components will be displayed.
- Rest of the procedure is same as for Pneumatic components discussed earlier.

P&ID COMPONENTS

The P&ID components are those components that are used in Piping and Instrumentation Diagrams. The procedure to insert the P&ID components is given next.

- Click on the **Insert P&ID Components** tool from the expanded **Insert Components** panel. The **Insert Component** dialog box with P&ID components will be displayed.
- Rest of the procedure is same as for Pneumatic components discussed earlier.

SUMMARY

- AutoCAD Electrical provides us a complete database of electrical components.
- This electrical component database has data like manufacturer, assembly code, catalog and so on that are used to identify the real-world match in the industry.
- When we generate the Bill of Material later, the catalog data is taken into account.
- To represent the components, we use schematic symbols which are available by Icon Menu tool. These symbols are mapped to the Component database available in the Catalog Browser.
- You can add the components in database by using the Catalog Browser.

Wires, Circuits, and Ladders

Chapter 5

Topics Covered

The major topics covered in this chapter are:

- *Inserting Wires*
- *Applying wire numbers*
- *Inserting user defined circuits*
- *Inserting ladders*
- *Cable Markers*
- *Circuit Builders*

INTRODUCTION

In the previous chapter, we learned to insert various electrical components. These components individually placed do not perform any task. But if connect them in a desired manner then various components perform various tasks. In this chapter, we will learn to add wires. Later, we will learn to create circuits, multiple buses, and ladders.

WIRES

Wires are the life line of any circuit. Wires are used to carry current and make the appliances run. The tools to create wires are available in the **Wire** drop-down in the **Insert Wires/Wire Numbers** panel in the **Schematic** tab of the **Ribbon**; refer to Figure-1.

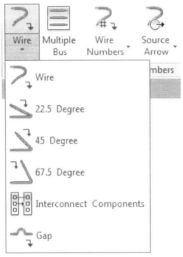

Figure-1. Wire drop-down

The tools in this drop-down are discussed next.

Wire

The **Wire** tool is used to add straight wires in the drawing to connect various components. The procedure to use this tool is given next.

- Click on the **Wire** tool from the **Wire** drop-down. The command prompt will be displayed as shown in Figure-2.

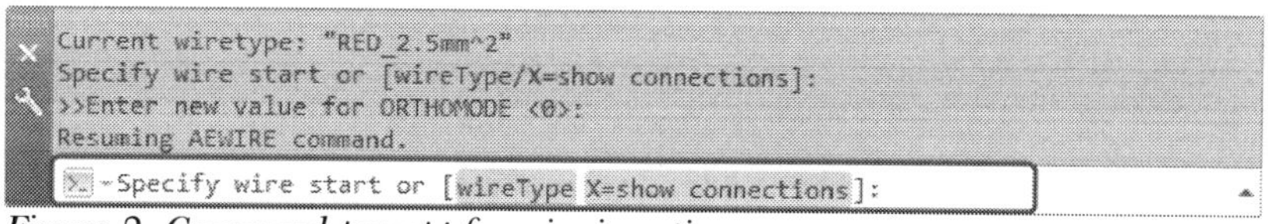

Figure-2. Command prompt for wire insertion

- Click in the **wireType** button in the command prompt or type **T** and press **ENTER** to change the wire type. The **Set Wire Type** dialog box will be displayed as shown in Figure-3.

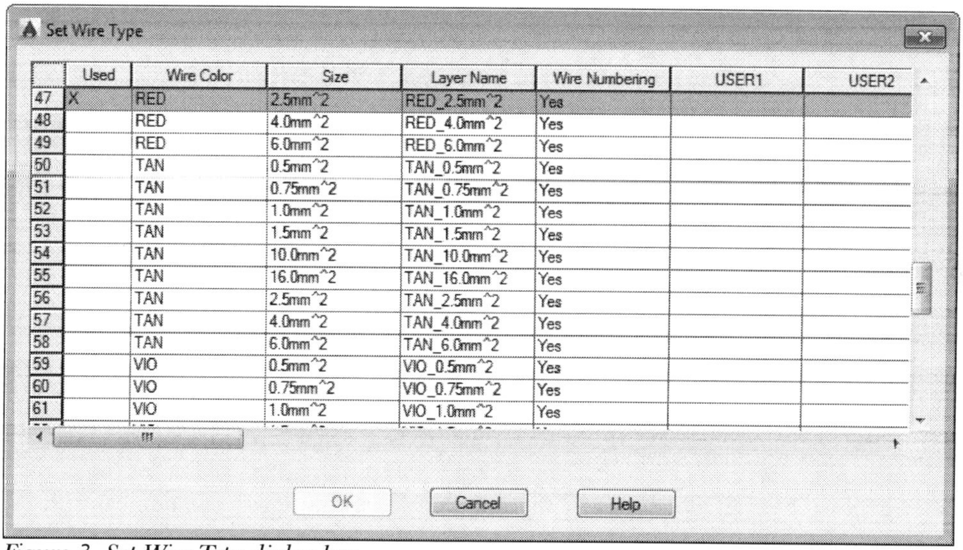

	Used	Wire Color	Size	Layer Name	Wire Numbering	USER1	USER2
47	X	RED	2.5mm^2	RED_2.5mm^2	Yes		
48		RED	4.0mm^2	RED_4.0mm^2	Yes		
49		RED	6.0mm^2	RED_6.0mm^2	Yes		
50		TAN	0.5mm^2	TAN_0.5mm^2	Yes		
51		TAN	0.75mm^2	TAN_0.75mm^2	Yes		
52		TAN	1.0mm^2	TAN_1.0mm^2	Yes		
53		TAN	1.5mm^2	TAN_1.5mm^2	Yes		
54		TAN	10.0mm^2	TAN_10.0mm^2	Yes		
55		TAN	16.0mm^2	TAN_16.0mm^2	Yes		
56		TAN	2.5mm^2	TAN_2.5mm^2	Yes		
57		TAN	4.0mm^2	TAN_4.0mm^2	Yes		
58		TAN	6.0mm^2	TAN_6.0mm^2	Yes		
59		VIO	0.5mm^2	VIO_0.5mm^2	Yes		
60		VIO	0.75mm^2	VIO_0.75mm^2	Yes		
61		VIO	1.0mm^2	VIO_1.0mm^2	Yes		

Figure-3. Set Wire Type dialog box

- Select the desired wire type from the list and click on the **OK** button from the dialog box. The selected wire will be set for insertion.
- Click in the drawing area to specify the start point of the wire. You will be asked to specify the end point of the wire and the command prompt will display as shown in Figure-4.

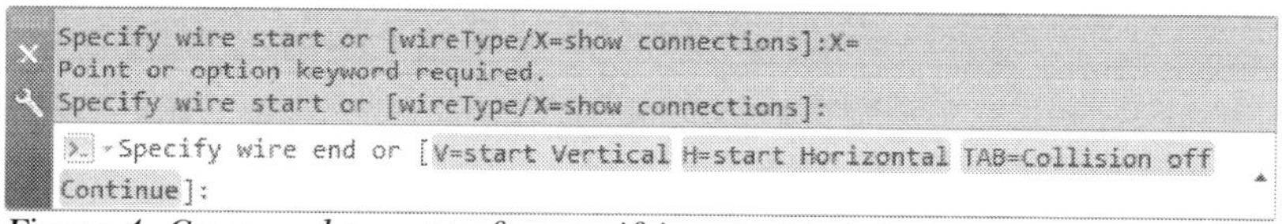

Figure-4. Command prompt after specifying start point

- If you are using the command prompt to specify the length of wire then you can use **V=start Vertical** or **H=start Horizontal** button in the command prompt to force AutoCAD Electrical draw only vertical or horizontal wires respectively.
- Click on the **Tab=Collision off** button to make the wires avoid collision with components. If you click again on this button then the wires will be able to pass through the components (at least in schematic drawings). Figure-5 shows two wirings one with collision OFF and one with collision ON.

Figure-5. Collision ON or OFF

- Click to specify the end point of the wire. You can go on creating the wires until you end up at a connector of a component. From there you need to start a new wire.

22.5 Degree, 45 Degree, and 67.5 Degree

These tools are used to create wires at their set angles. The procedure to use these tools is given next.

- Click on the **22.5 Degree** or **45 Degree** or **67.5 Degree** tool from the **Wire** drop-down. You will be prompted to specify the starting point of the wire.
- Click at the desired point to start wire. You are asked to specify the end point of the wire.
- Click to specify the end point of the wire.

Interconnect Components

The **Interconnect Components** tool is used to interconnect two components with each other. The procedure to use this tool is given next.

- Click on the **Interconnect Components** tool from the **Wire** drop-down. You are asked to select the first component.
- Select the first component. You will be prompted to select the second component.
- Select the second component. The wire will be created connecting both the components; refer to Figure-6.

Note that the components to be connected using this tool should be aligned properly so that there connection points are in same orientation.

Figure-6. Components interconnected

Gap

This tool is used to create gap in the wiring where the two wires intersect each other. The procedure to use this tool is given next.

- Click on the **Gap** tool from the **Wire** drop-down. You are asked to select the wire that you want to be remained solid while applying this tool.
- Select the desired wire. You are asked to select the crossing wire.
- Select the intersecting wire. Gap in the wires will be created; refer to Figure-7.

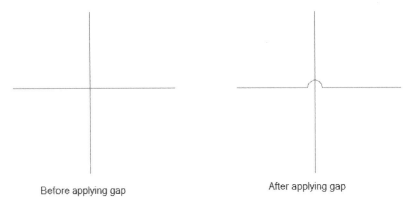

Figure-7. Wires on using Gap tool

MULTIPLE BUS

The **Multiple Bus** tool is used to create multiple lines of wires. To create a multiple bus, follow the steps given next.

- Click on the **Multiple Bus** tool from the **Insert Wires/Wire Numbers** drop-down. The **Multiple Wire Bus** dialog box will be displayed as shown in Figure-8.
- Click in the Spacing edit boxes so specify the horizontal and vertical spacing between the wires.
- By default, the Another Bus (Multiple Wires) radio button is selected. As a result, you can create the multiple wires from a point on another wire.
- Select the **Empty Space, Go Horizontal** radio button or the **Empty Space, Go Vertical** radio button to create multiple wires in horizontal or vertical direction respectively.

Creating of Multiple Buses with various radio button

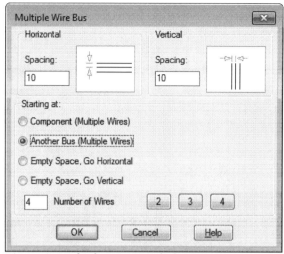

Figure-8. Multiple Wire Bus dialog box

Component (Multiple Wires)

- Select the **Component (Multiple Wires)** radio button from the **Multiple Wire Bus** dialog box.
- Click on the **OK** button from the dialog box. You are prompted to create a selection window to select the connection ports of devices.
- Select the connection ports that you want to use for creating multiple wire bus; refer to Figure-9. The selected ports will be highlighted as shown in Figure-10.

Figure-9. Window selection of ports

Figure-10. Connection ports highlighted

- Press ENTER at the command prompt. Wires will get attached to the cursor; refer to Figure-11.

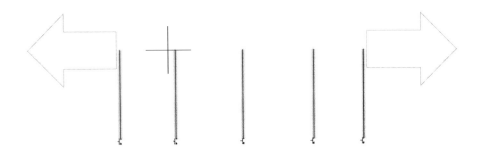

Figure-11. Wires attached to the cursor

- Click to specify the length of the wires. The multiple wire bus will be created.

Another Bus (Multiple Wires)

The **Another Bus (Multiple Wires)** radio button is used to create a multiple wire bus by using an already existing wire bus. The steps to use this option are given next.

- Select the **Another Bus (Multiple Wires)** radio button from the **Multiple Wire Bus** dialog box. Specify the number of wires in the **Number of Wires** edit box. Note that the number of wires should be equal to the wires in the already existing bus.
- Click on the **OK** button from the dialog box. You are asked to select a wire.
- Select the middle wire from the already existing wire bus. The ends of the current wire bus will get attached to the cursor; refer to Figure-12.

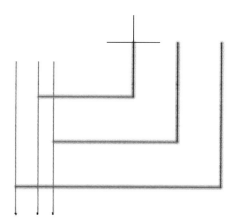

Figure-12. Wire bus from another wire bus

- Click to specify the end point of the wire bus.

Empty Space, Go Horizontal

The **Empty Space, Go Horizontal** radio button is used to create a multiple wire bus in horizontal direction starting from the specified start point. The steps to use this option are given next.

- Select the **Empty Space, Go Horizontal** radio button from the **Multiple Wire Bus** dialog box. Specify the number of wires in the **Number of Wires** edit box.
- Click on the **OK** button from the dialog box. You are asked to specify the starting point for the bus.
- Click in the drawing area. You are asked to specify the end point of the bus.
- Click to specify the end point.

Empty Space, Go Vertical

The **Empty Space, Go Vertical** radio button is used to create a multiple wire bus in vertical direction starting from the specified start point. The steps to use this option are given next.

- Select the **Empty Space, Go Vertical** radio button from the **Multiple Wire Bus** dialog box. Specify the number of wires in the **Number of Wires** edit box.
- Click on the **OK** button from the dialog box. You are asked to specify the starting point for the bus.
- Click in the drawing area. You are asked to specify the end point of the bus.
- Click to specify the end point.

LADDERS

In Electrical systems, multiple circuits that are powered by a common power source can be combined in the form of ladders. Ladders are arrangement of wire as shown in Figure-13.

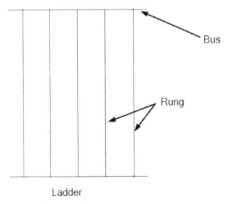

Figure-13. Ladder

Using these Ladders, the Ladder Diagrams are created which are backbone to PLCs in today's world. The tools to create and control ladders are available in the **Ladder** drop-down; refer to Figure-14. The procedures to use the tools in this drop-down are given next.

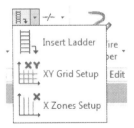

Figure-14. Ladder drop down

Insert Ladder

The **Insert Ladder** tool is used to insert the ladder of wires in the drawing. The procedure to use this tool is given next.

- Click on the **Insert Ladder** tool in the **Ladder** drop-down. The **Sheet 1: Insert Ladder** dialog box will be displayed as shown in Figure-15.

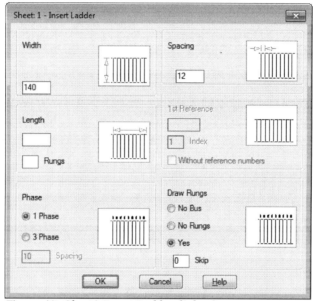

Figure-15. Sheet1 Insert Ladder dialog box

- Click in the edit box in **Width** area and specify the distance between two wire buses.
- Click in the edit box in **Spacing** area and specify the distance between two rungs.
- Click in the **Length** edit box and specify the total length of the ladder or click in the **Rungs** edit box and specify the number of rungs in the ladder. Note that you can specify the value in any one of the two edit boxes. Value in the other edit boxes will be calculated automatically.
- Select the desire number of phases for the ladder. If you select the **3 Phase** radio button from the **Phase** area then the ladder will be created with three phase wire lines and a value of distance between phase lines will be required in the **Spacing** edit box.

- You can skip the creation of bus or rung in the ladder by selecting the respective radio button from the **Draw Rungs** area of the dialog box.
- Click on the **OK** button from the dialog box to create the ladder.

XY Grid Setup

This tool is used to setup a grid of wires based on the specified values. This grid can be later used to connect various circuits. The procedure to use this tool is given next.

- Click on the **XY Grid Setup** tool from the **Ladder** drop-down. The **X-Y Grid Setup** dialog box will be displayed as shown in Figure-16.

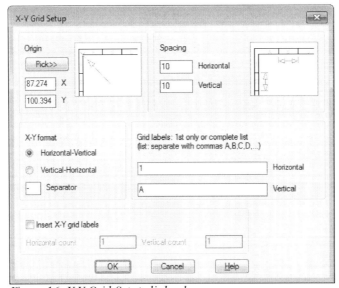

Figure-16. X-Y Grid Setup dialog box

- Click on the **Pick>>** button from the dialog box to specify the origin of the grid. You are prompted to pick a point in the drawing area.
- Click in the drawing area to specify the point.
- Specify the spacing between horizontal and vertical grid lines in the edit boxes available in the Spacing area of the dialog box.

- Select the desired radio button from the **X-Y format** area of the dialog box.
- Select the **Insert X-Y grid labels** check box to display the labels of the grid.
- After selecting the check box specify the horizontal and vertical counts in the respective edit boxes below the check box.
- Click on the **OK** button from the dialog box. The grid will be created; refer to Figure-17. Note that the grid created is for reference purpose. Using this grid, you can precisely place the components in the drawing.

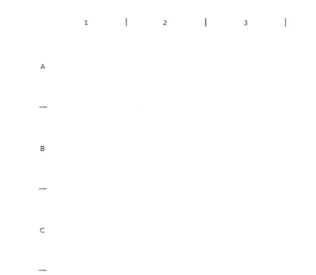

Figure-17. Grid created

X Zones Setup

The **X Zones Setup** tool is also used to create vertical references in the drawing for inserting components. Note that you can use either the **X-Y Grid Setup** tool or **X Zone Setup** tool. Before creating X zones, you need to enable this option. To enable this option, follow the steps given next.

- Right-click on the name of current drawing in the **Project Manager**. A shortcut menu will be displayed as shown in Figure-18.

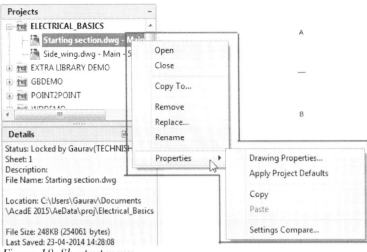
Figure-18. Shortcut menu

- Move the cursor to **Properties** option in the shortcut menu. A cascading menu will be displayed.
- Click on the **Drawing Properties** option from the menu. The **Drawing Properties** dialog box will be displayed as shown in Figure-19.

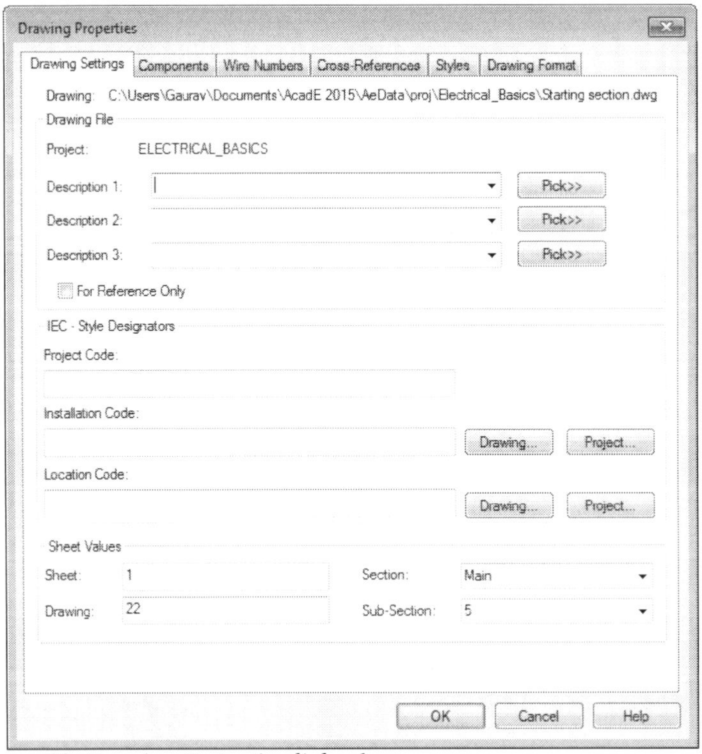

Figure-19. Drawing Properties dialog box

- Click on the Drawing Format tab in the dialog box. The options in the dialog box will be displayed as shown in Figure-20.

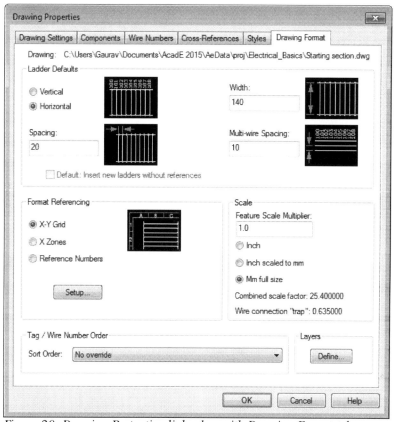

Figure-20. Drawing Properties dialog box with Drawing Format tab

- Select the X Zones radio button from the **Format Referencing** area of the dialog box.
- Note that you can also define the default direction of ladder creation by selecting the desired radio button from the **Ladder Defaults** area of the dialog box.
- Click on the OK button from the dialog box to apply the specified settings.

Now, we have enabled the **X Zones Setup** tool and we can use it in our drawing. The procedure to use the **X Zones Setup** tool is given next.

- Click on the **X Zones Setup** tool from the **Ladder** drop-down in the **Insert Wires/Wire Numbers** panel in the **Ribbon**. The

X Zones Setup dialog box will be displayed as shown in Figure-21.

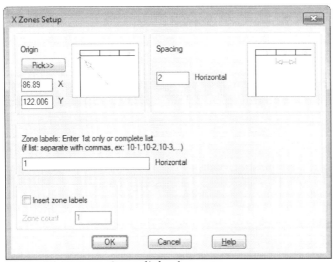

Figure-21. X Zones Setup dialog box

- Click on the **Pick>>** button from the **Origin** area of the dialog box. You are prompted to select a point in the drawing area to specify the origin.
- Click on the desired point in the drawing area.
- Click in the **Horizontal** edit box in the **Spacing** area of the dialog box and specify the desired distance value.
- Select the Insert zone labels check box to include the labels on the zone.
- Click on the **OK** button from the dialog box. The X zone will be created; refer to Figure-22.

Figure-22. X zones

When we work on real world projects of electrical systems, then we generally require references to insert our components. At the time, XY grids and X Zones are required to create references.

WIRE NUMBERING

Wire numbering is an important aspect of wires. Wire number helps to identify circuits related to that wire, components connected to that wire, total length of the wire and other details. The tools to specify wire numbers are available in the **Wire Numbers** drop-down in the **Insert Wires/Wire Numbers** panel. There are three tools in this drop-down; **Wire Numbers**, **3 Phase**, and **PLC I/O**. Refer to Figure-23. These tools are discussed next.

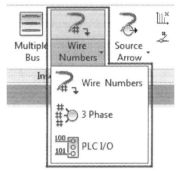

Figure-23. Wire Numbers drop-down

Wire Numbers

This tool is used to specify the wire number for individual wires or project wide. The procedure to use this tool is given next.

- Click on the **Wire Numbers** tool from the **Wire Numbers** drop-down. The **Sheet1 - Wire Tagging** dialog box will be displayed as shown in Figure-24.

Figure-24. Sheet1 Wire Tagging dialog box

- Click in the **Start** edit box in the **Wire tag mode** area and specify the starting number for wire tagging(numbering).
- Select the **Tag new/un-numbered only** radio button if you want to specify the wire number to un-numbered wires only otherwise leave on the default.
- Click on the **Setup** button from the dialog box to specify the format of wire numbering. The **Assign Wire Numbering Formats by Wire Layers** dialog box will be displayed as shown in Figure-25.

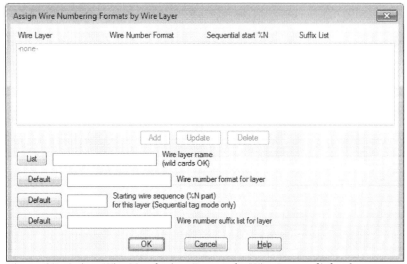

Figure-25. Assign Wire Numbering Formats by Wire Layers dialog box

- Click on the **List** button in this dialog box. The **Select Wire Layer** dialog box will be displayed; refer to Figure-26.

Figure-26. Select Wire Layer dialog box

- Select the desired color layer for the wire.
- Click on the **OK** button from the **Select Wire Layer** dialog

box.
- Click on the three **Default** buttons one by one to specify the default values in the related fields.
- Click on the **Add** button from the dialog box, the wire numbering format will be added.
- Click on the **OK** button from the dialog box.
- Now, we are ready to use this layer format in our wire numbering. So, select the Use wire layer format override check box.
- If you want to apply the wire numbering individually to each wire then click on the **Pick Individual Wires** button from the dialog box. You are prompted to select the wires.
- Select the wires one by one and press ENTER. The numbering will be assigned to the wires; refer to Figure-27.

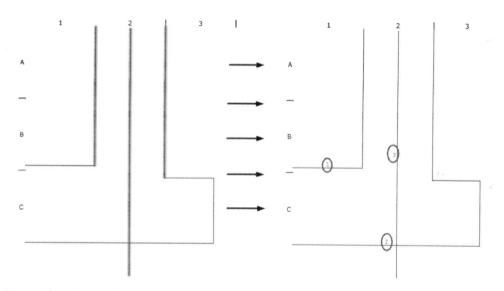

Figure-27. Wire numbering

3 Phase

This tool is used to number three phase wiring. The procedure to use this tool is given next.

- Click on the **3 Phase** tool from the **Wire Numbers** drop-down in the **Ribbon**. The **3 Phase Wire Numbering** dialog box will be displayed as shown in Figure-28.

Figure-28. 3 Phase Wire Numbering dialog box

- Specify the desired values in the **Prefix, Base,** and **Suffix** edit boxes.
- Select the desired radio button from the **Maximum** area of the dialog box. If you have 3 wires in the connection then select **3** radio button and if you have 4 then select the **4** radio button.
- Click on the **OK** button from the dialog box. You are asked to select first wire of the 3 phase line.
- Select the respective wire. You are asked to select the second wire.
- Select the second wire and similarly select the third wire.
- After selecting the third wire, the dialog box will be displayed again. Keep on selecting the individual wire that you want to be numbers.
- After numbering, click on the **Cancel** button from the dialog box to exit.

PLC I/O

This tool is used to number the input and output ports of the PLC. The procedure to use this tool is given next.

- Click on the **PLC I/O** tool from the **Wire Numbers** drop-down in the **Ribbon**. The **PLC I/O Wire Numbers** dialog box will be displayed as shown in Figure-29.

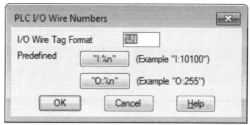

Figure-29. PLC I O Wire Numbers dialog box

- Specify the desired tag format for the PLC I/O. The % codes have already been discussed.
- Click on the **OK** button from the dialog box. You are prompted to select the PLC.
- Select a PLC, you are asked to select the wires connected to the PLC.
- One by one select the wires that are connected to the PLC.
- The tags defined as per the PLC are extracted and applied to the wires; refer to Figure-30.

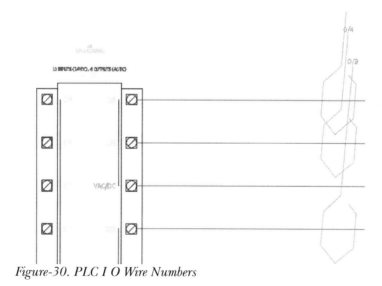

Figure-30. PLC I O Wire Numbers

WIRE NUMBER LEADERS AND LABELS

In the previous figure, the wire number leaders are assigned automatically. We can modify the wire number leaders and we can assign the labels as per our need. The tools to do so are available in the **Wire Number Leader** drop-down; refer to Figure-31.

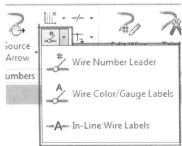

Figure-31. Wire Number Leader drop down

The procedure to use the tools in this drop-down is discussed next.

Wire Number Leader

The Wire Number Leader tool is used to assign leader to a wire number. The procedure to use this tool is given next.

- Click on the **Wire Number Leader** tool from the **Wire Number Leader** drop-down in the **Ribbon**. You will be prompted to select a wire number; refer to Figure-32.

Figure-32. Command prompt for wire numbers

- Select the desired wire number from the drawing. You will be prompted to specify the point for leader end point; refer to Figure-33.

Figure-33. After selecting wire number

- Click on the drawing area to specify the end point of the wire number leader. You will be prompted to specify the next point of the leader.
- Click in the drawing area to specify the point or press ENTER. The wire number leader will be assigned; refer to Figure-34.

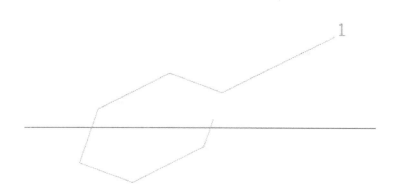

Figure-34. Wire number leader assigned

- Similarly, you can assign leaders to other wire numbers.

Wire Color/Gauge Labels

This tool is used to attach the wire color and/or gauge labels to the wires so that we can easily identify the wires in the circuit. The procedure to use this tool is given next.

- Click on the **Wire Color/Gauge Labels** tool from the **Wire Number Leader** drop-down in the **Ribbon**. The **Insert Wire Color/Gauge Labels** dialog box will be displayed as shown in Figure-35.
- Click on the **Setup** button to specify the settings. The **Wire label color/gauge setup** dialog box will be displayed as shown in Figure-36.

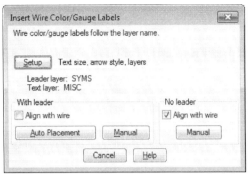

Figure-35. Insert Wire Color or Gauge Labels dialog box

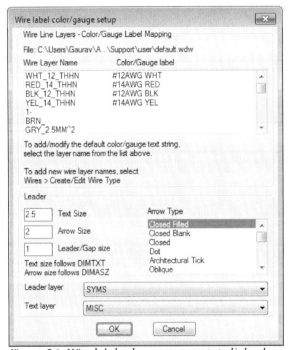

Figure-36. Wire label color or gauge setup dialog box

- Specify the desired settings in the dialog box and click on the **OK** button from the dialog box.
- Click on the **Manual** button from the dialog box to manually assign the color/gauge labels. Note that you can assign the labels with leaders or without leaders by selecting the **Manual** button from the respective area.
- Click on the **Auto Placement** button to automatically place the color/gauge labels.
- After selecting the desired button, click on the wire

and follow the instructions given by system. Press ENTER to create the labels. The labels will be attached to the wires; refer to Figure-37.

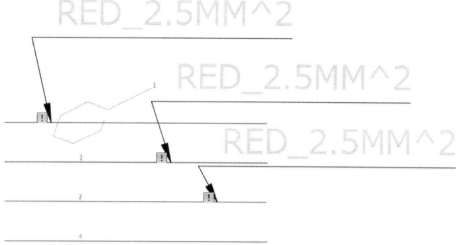

Figure-37. Wire labels

In-Line Wire Labels

The previous tool is used to insert labels aside the wires. The In-Line Wire Labels tool is used to insert the in-line wire labels. The labels can include informations like color, gauge, and so on. The procedure to use this tool is given next.

- Click on the **In-Line Wire Labels** tool from the **Wire Number Leader** drop-down. The **Insert Component** dialog box will be displayed; refer to Figure-38.

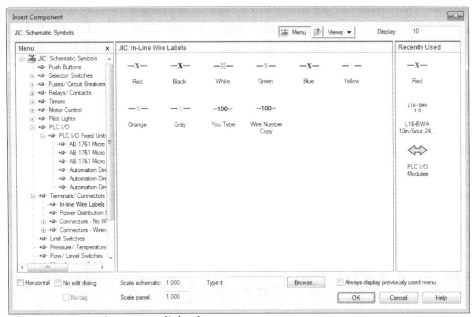

Figure-38. Insert Component dialog box

- Select the desired symbol from the dialog box. You will be prompted to specify the insertion point for the symbol.
- Click on the wire to insert the label. The label will be inserted; refer to Figure-39.

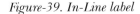

Figure-39. In-Line label

MARKERS

The markers are used to mark the wires for categorizing them. The tools to apply marks are available in the **Cable Markers** and **Insert Dot Tee Markers** drop-down. The procedure to use all these tools are similar. These procedures are discussed next.

Cable Markers

This tool is used to insert cable markers. The procedure is given next.

- Click on the **Cable Markers** tool from the **Cable Markers** drop-down. The **Insert Component** dialog box will be displayed with the cable markers; refer to Figure-40.

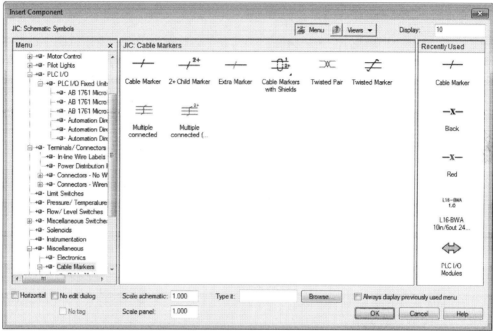

Figure-40. Insert Component dialog box with cable markers

- Click on the desire symbol and then click on the wire to place the symbol. The **Insert/Edit Cable Marker (Parent wire)** dialog box will be displayed as shown in Figure-41.

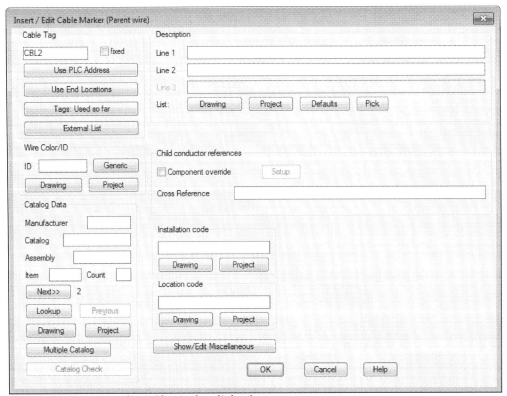

Figure-41. Insert or Edit Cable Marker dialog box

- Specify the parameters as desired and click on the **OK** button from the dialog box. The **Insert Some Child Components** dialog box will be displayed; refer to Figure-42.

Figure-42. Insert Some Child Components dialog box

- If you want to insert child components then click on the **OK Insert Child** button otherwise click on the **Close** button from the dialog box.
- If you click on the **OK Insert Child** button then you need to click on the child wires and specify the tags as desired.

Multiple Cable Markers

The **Multiple Cable Markers** tool is used to insert multiple cable markers in the drawing. The procedure to use this tool is given next.

- Click on the **Multiple Cable Markers** tool from the **Cable Markers** drop-down. The **Multiple Cable Markers** dialog box will be displayed as shown in Figure-43.

Figure-43. Multiple Cable Markers dialog box

- Select the desired radio button from the dialog box and click on the **OK** button from the dialog box. The cable markers will be inserted automatically.

Insert Dot Tee Markers

The **Insert Dot Tee Markers** tool is used to insert dot tee mark on the wire. The procedure to use this tool is given next.

- Click on the **Insert Dot Tee Markers** tool from the **Insert Dot Tee Markers** drop-down in the **Ribbon**. You will be prompted to click on the wire to create the dot tee mark.
- Click on the wire at desired point. The dot tee mark will be created; refer to Figure-44.

Figure-44. Dot tee mark

Insert Angled Tee Markers

The **Insert Angled Tee Markers** tool is used to convert the tee mark into an angles tee mark. The procedure to use this tool is given next.

- Click on the **Insert Angled Tee Markers** tool from the **Insert Dot Tee Markers** drop-down. You will be prompted to select the tee of wires.
- Click on the tee, it will be converted to angled tee mark; refer to Figure-45.

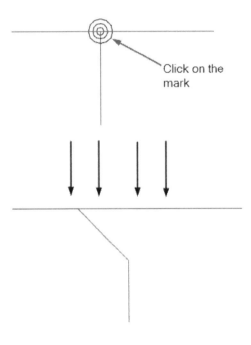

Figure-45. Angled tee

CIRCUIT BUILDER

The **Circuit Builder** tool is used to insert circuits in the drawing. The procedure to use this tool is given next.

- Click on the **Circuit Builder** tool from the **Circuit Builder** drop-down in the **Insert Components** panel in the **Ribbon**; refer to Figure-46. The Circuit Selection dialog box will be displayed as shown in Figure-47.

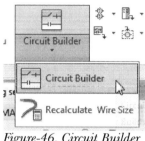

Figure-46. Circuit Builder tool

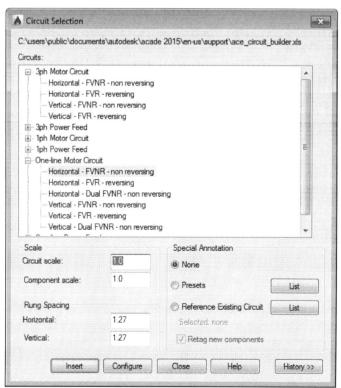

Figure-47. Circuit Selection dialog box

- Select the desired circuit from the list.

- Click in the edit boxes of **Scale** area and specify the scale values for circuits and components.
- Specify the spacing between rungs in the **Rung Spacing** area.
- Select the desired radio button from the **Special Annotation** area of the dialog box.
- Click on the **Insert** button from the dialog box. You will be prompted to specify the insertion point of the circuit.
- Click to specify the insertion point, the circuit will be placed; refer to Figure-48.

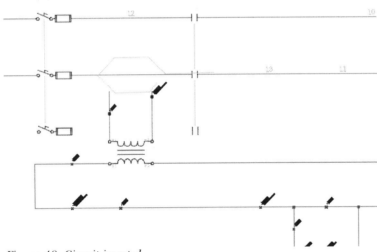

Figure-48. Circuit inserted

PRACTICAL

In this practical, we will create the circuit diagram of an electrical system; refer to Figure-49.

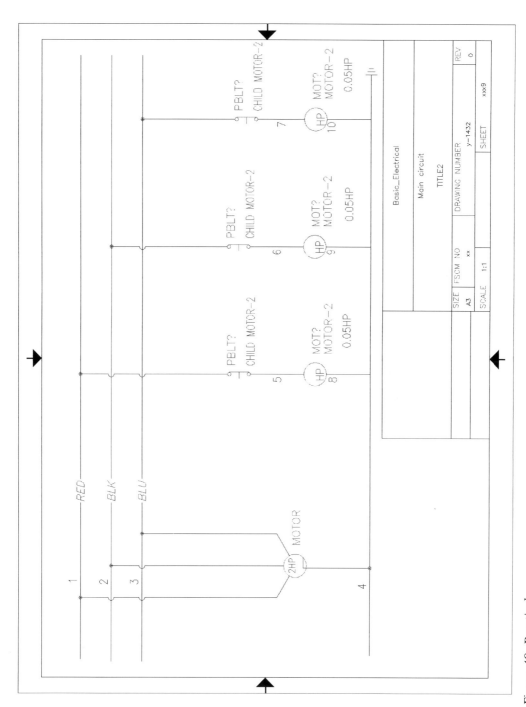

Figure-49. Practical

Starting a New drawing

- Start AutoCAD Electrical and Click on the down arrow below Start Drawing button. A list of templates will be displayed; refer to Figure-50.

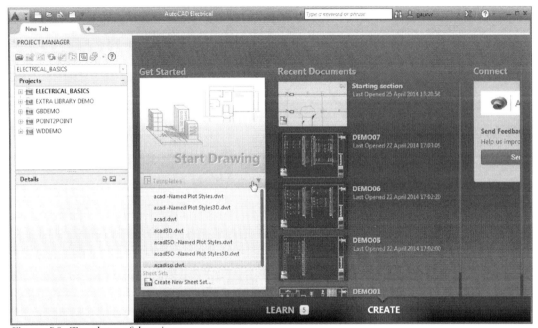

Figure-50. Templates of drawings

- Click on the **ACE ANSI A(Landscape) Color.dwt** template from the list of templates. A new drawing will open with the selected template; refer to Figure-51.

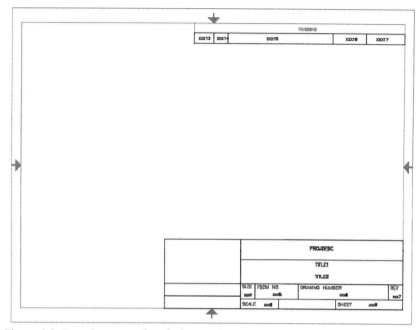
Figure-51. Drawing opened with the template

Editing Title Block

- Double-click on the title block at the bottom right corner. The **Enhanced Attribute Editor** dialog box will be displayed; refer to Figure-52.

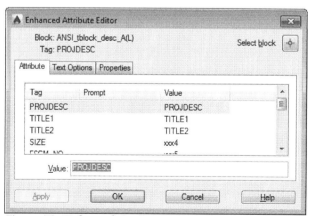
Figure-52. Enhanced Attribute Editor dialog box

- Select the desired tag from the list and specify the respective value.

- Click on the **OK** button from the dialog box to exit.

Creating Wires

- Click on the **Multiple Bus** tool from the **Insert Wires/Wire Numbers** panel in the **Ribbon**. The **Multiple Wire Bus** dialog box will be displayed as shown in Figure-53.

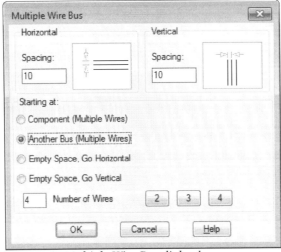

Figure-53. Multiple Wire Bus dialog box

- Specify the spacing as **0.5** in both the **Spacing** edit boxes.
- Select the **Empty Space, Go Horizontal** radio button from the dialog box.
- Click on the **3** button to specify the number of wires as 3.
- Click on the **OK** button. You are asked to specify the starting point of the wire bus.
- Click on the drawing area and create the wire bus as shown in Figure-54.

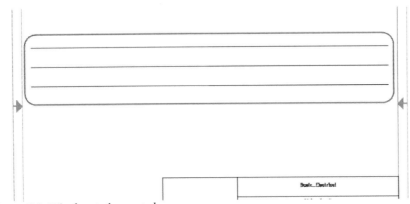

Figure-54. Wire bus to be created

- Similarly, create two more buses each having 3 wires with proper spacing; refer to Figure-55.

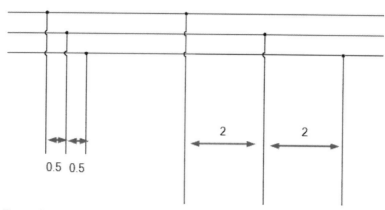

Figure-55. Other two wire buses

- Click on the **Wire** tool from **Insert Wires/Wire Numbers** panel and create a wire for earth; refer to Figure-56.

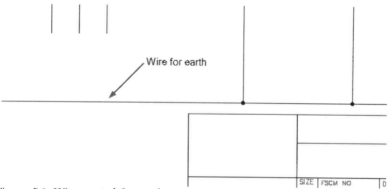

Figure-56. Wire created for earth

Assigning Numbers and Labels to Wires

- Click on the **In-Line Wire Labels** tool from the **Wire Number Leader** drop-down; refer to Figure-57. The Insert Component dialog box with wire labels will be displayed; refer to Figure-58.

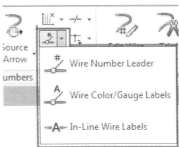

Figure-57. Wire Number Leader drop down

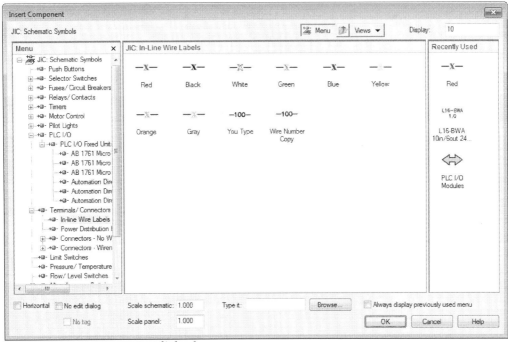

Figure-58. Insert Component dialog box

- Select the Red symbol from the dialog box and click on the top wire to specify its color code.
- Similarly, set the middle and bottom wire as **Black** and **Blue** respectively; refer to Figure-59. Note that by default, the tags on wires are displayed in yellow color. To change the color, double click on the color name in the label. The **Enhanced Attribute Editor** dialog box will be displayed; refer to Figure-60.

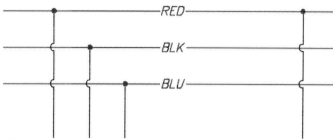

Figure-59. Wires after color coding

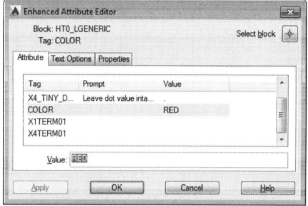

Figure-60. Enhanced Attribute Editor dialog box for color

- Click on the **Properties** tab and change the color of layer as per your requirement.
- Click on the **OK** button from the dialog box.

- Click on the **Wire Numbers** tool from the **Wire Numbers** drop-down. The **Sheet1 - Wire Tagging** dialog box will be displayed as shown in Figure-61.

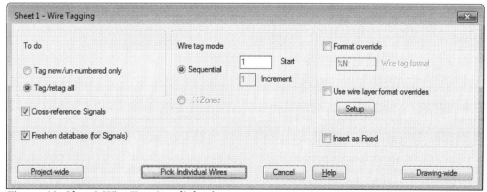

Figure-61. Sheet1 Wire Tagging dialog box

- Click on the **Drawing-wide** button from the dialog box. The wire numbers will be assigned to the wires; refer to Figure-62.

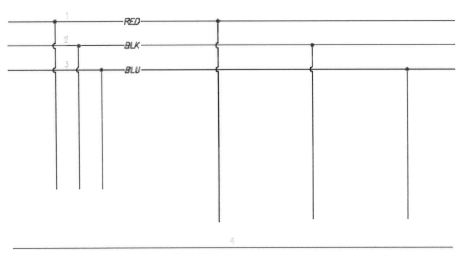

Figure-62. Wire numbers assigned to the wires

- You can change the color of wire numbers as we did for the wire labels.

Inserting 3 Phase Motor

- Click on the **Catalog Browser** button from the **Components** drop-down; refer to Figure-63. The **CATALOG BROWSER** will be displayed.
- Click in the **Category** drop-down and select the **MO (Motors)** option from the list.
- Click in the **Search** edit box and specify **4-Pole** in it.
- Click on the **Search** button. The list of motors in the catalog will be displayed as shown in Figure-64.

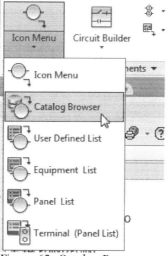

Figure-63. Catalog Browser button

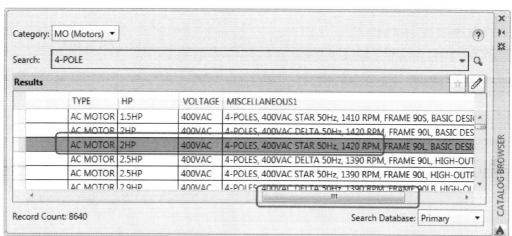

Figure-64. Catalog Browser with list of motors

- Select the Motor as highlighted in the red box in above figure. The panel of buttons will be displayed; refer to Figure-65.

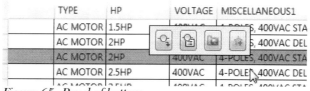

Figure-65. Panel of buttons

- Click on the button. The Insert Component dialog box

will be displayed; refer to Figure-66.

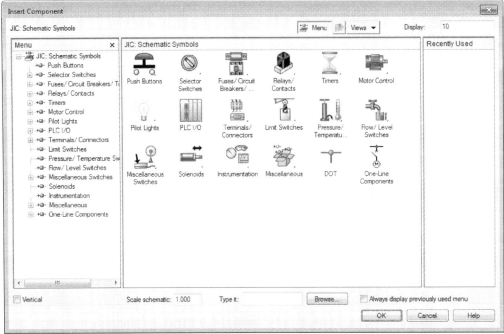

Figure-66. Insert Component dialog box by default

- Select the **Motor Control** category from the menu and select the icon highlighted by the cursor; refer to Figure-67.

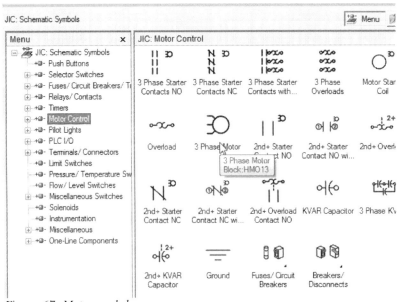

Figure-67. Motor symbol

- On doing so, the **Assign Symbol To Catalog Number** dialog box will be displayed; refer to Figure-68.

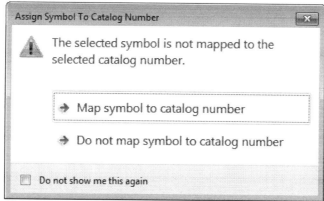

Figure-68. Assign Symbol To Catalog Number dialog box

- Click on the **Map symbol to catalog number** button. The symbol will get attached to cursor.
- Click on the middle wire as shown in Figure-69. The motor will get attached to the three wires automatically and the **Insert / Edit Component** dialog box will be displayed as shown in Figure-70.

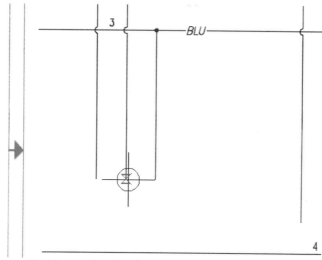

Figure-69. Motor to be placed

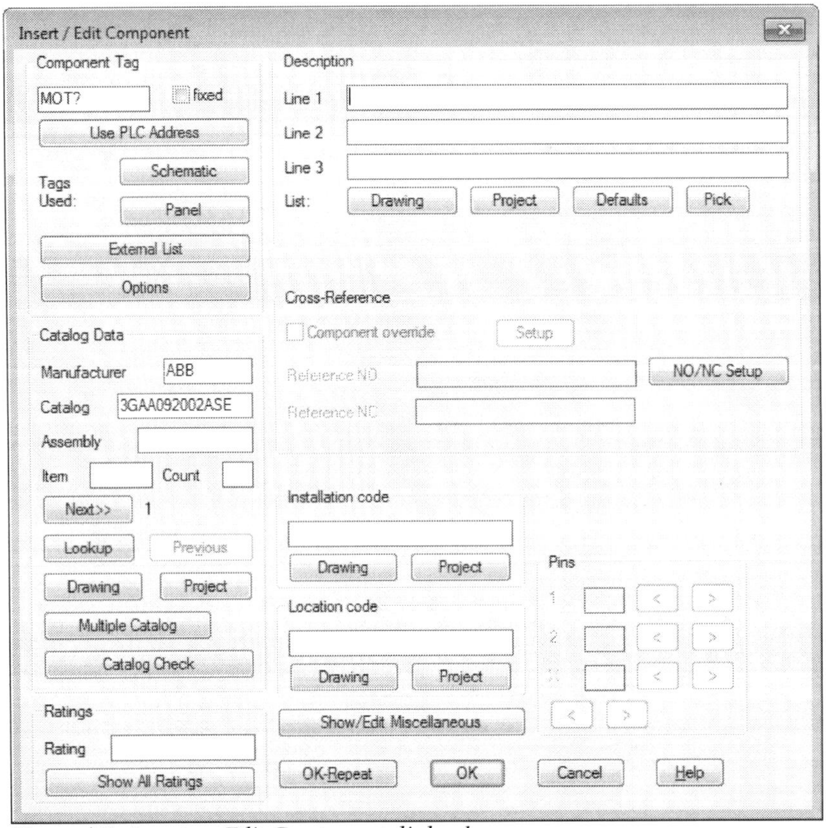

Figure-70. Insert or Edit Component dialog box

- Click in the edit box in **Component Tag** area and specify the tag as **Motor**.
- Click in the **Line 1** edit box and specify description as **3 Phase**.
- Click in the **Rating** edit box in **Ratings** area and specify the rating as **2 HP**.
- Click on the **OK** button from the dialog box. The motor will be placed; refer to Figure-71.

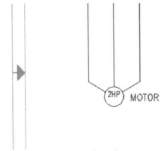

Figure-71. Motor placed

Adding Ground symbol

- Click on the **Icon Menu** button from the **Icon Menu** drop-down. The **Insert Component** dialog box will be displayed.
- Select the **Motor Control** category from the **Menu** area and click on the **Ground** symbol from the right area of the dialog box; refer to Figure-72. You are prompted to specify the location of the ground symbol.

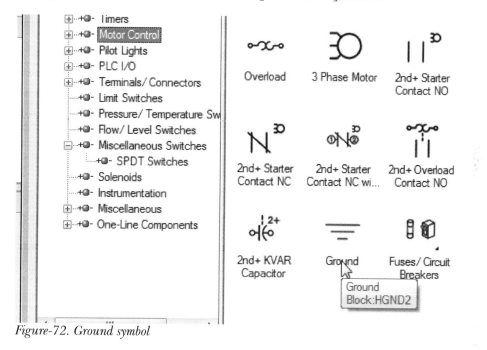

Figure-72. Ground symbol

- Click at the end of wire with 4 wire number; refer to Figure-73.

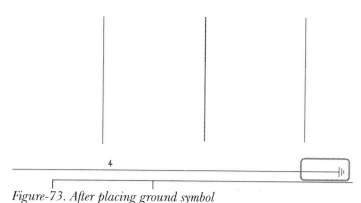

Figure-73. After placing ground symbol

Adding symbols for various components

- Click on the **Catalog Browser** button and insert the components as shown in Figure-74.

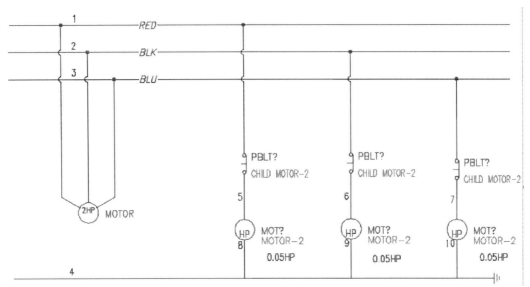

Figure-74. Circuit after connecting components

- Connect the **2 HP Motor** to the ground wire using the **Wire** tool; refer to Figure-75.

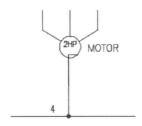

Figure-75. Wire connected to motor

- After adding all the components, the drawing will be displayed as shown in Figure-76.

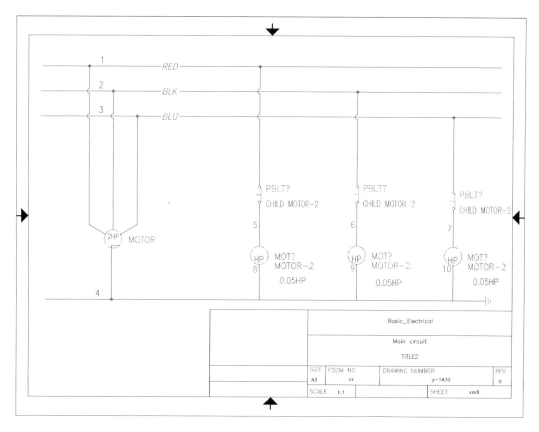

Figure-76. Practical-Model

Problem

Create the circuit diagram as shown in Figure-77.

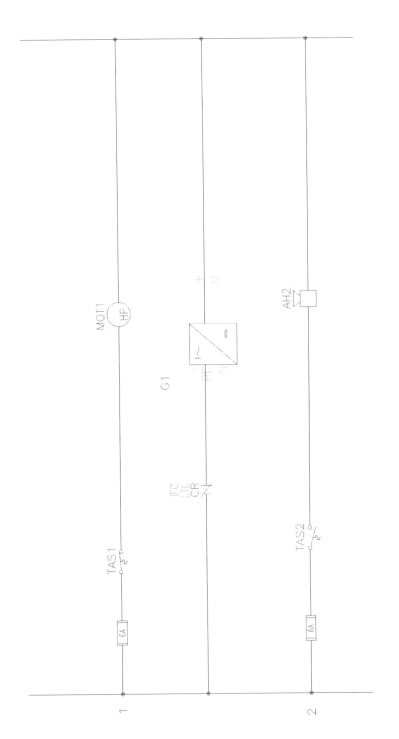

Figure-77. Problem-Model

Editing Wires, Components, and Circuits

Chapter 6

The major topics covered in this chapter are:

- **Editing Components**
- **Internal Jumpers**
- **Editing Components like fixing tags, copying catalog assignments**
- **Editing Circuits**
- **Editing Wires and Wire Numbers**

INTRODUCTION

Till this point, we have learned to insert components, insert wires, and insert circuits. Now, we are ready to edit the components, wires and circuits for customizing. The tools to edit components are available in three panels; **Edit Components**, **Circuit Clipboard**, and **Edit Wires/Wire Numbers** panel; refer to Figure-1, Figure-2, and Figure-3.

Figure-1. Edit Components panel

Figure-2. Circuit Clipboard panel

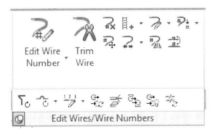

Figure-3. Edit Wires or Wire Numbers panel

EDIT TOOL

The **Edit** tool is available in the **Edit Components** drop-down in the **Edit Components** panel; refer to Figure-4. This tool is used to edit the details of any component inserted in the drawing area. The procedure to use this tool is given next.

Figure-4. Edit components drop-down

- Click on the **Edit** tool from the **Edit Components** drop-down. You will be prompted to select the components/cables/location boxes.
- Select the component. The **Insert/Edit Component** dialog box will be displayed; refer to Figure-5. The options in this dialog box have already been discussed.
- Specify the desired settings in the dialog box and click on the **OK** button from the dialog box.
- If you want to use the component with the specified settings again then click on the **OK-Repeat** button from the dialog box.

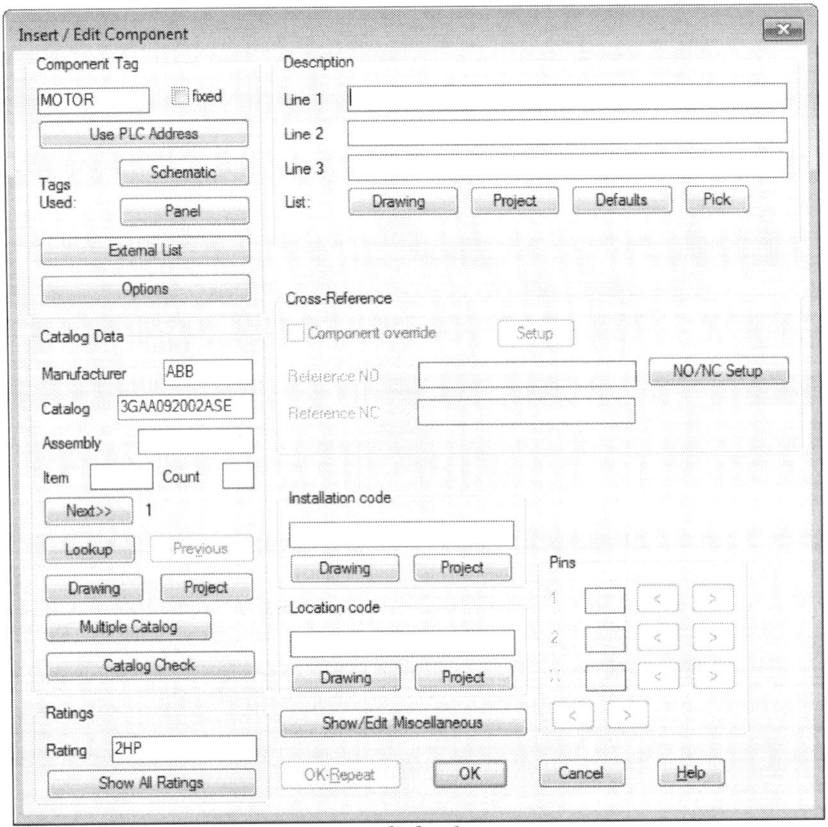

Figure-5. Insert or Edit Component dialog box

INTERNAL JUMPER

The Internal Jumper is used to interconnect terminals of an electrical components. A few components in electrical system have jumpers to change their functionality. For example, you might have seen jumpers in transformers to change the output voltage. The **Internal Jumper** tool in the **Edit Components** drop-down is used to edit the functions of internal jumpers. The procedure to use this tool is given next.

- Click on the **Internal Jumper** tool from the **Edit Components** drop-down. You are asked to select a component for editing jumpers.
- Select the component; refer to Figure-6. The **Wire Jumpers** dialog box will be displayed; refer to Figure-7.

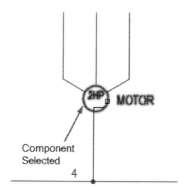

Figure-6. Component selected

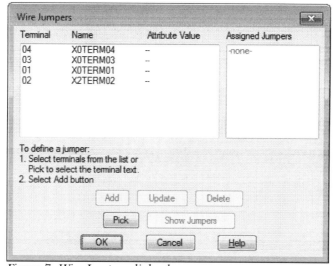

Figure-7. Wire Jumpers dialog box

- Click on the **Pick** button from the dialog box. You are asked to select the jumper points on the component; refer to Figure-8. Note that you can select the terminals directly from the dialog box but in that case you will not be able to identify the wire connections to the jumpers.
- Press ENTER. The **Wire Jumpers** dialog box will be displayed with the jumpers selected; refer to Figure-9. Note that the **Add** button is active now.

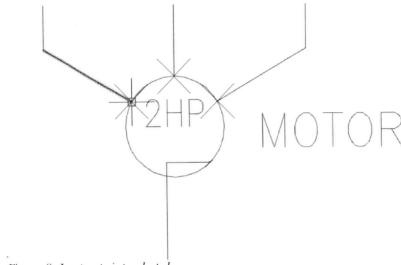

Figure-8. Jumper points selected

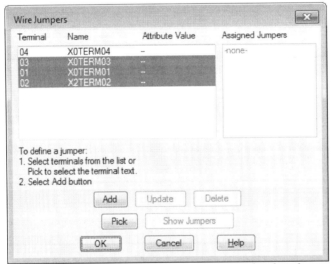

Figure-9. Wire Jumpers dialog box with terminals selected

- Click on the **Add** button from the dialog box. The jumper will be connected to the selected terminals and will be listed in the **Assigned Jumpers** area of the dialog box; refer to Figure-10.

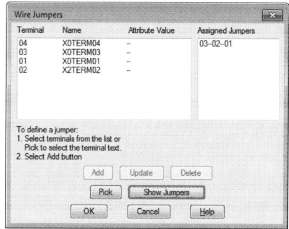

Figure-10. Wire Jumpers dialog box with assigned jumpers

- Click on the Show Jumpers button to display the jumper in the drawing area. The jumper will be displayed connecting all the selected terminals; refer to Figure-11.

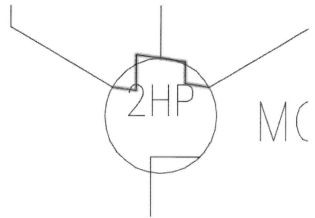

Figure-11. Jumper connecting the terminals

The example use above is just for illustration purpose, in real world we will not be connecting motors like this. We can use PLCs for applying jumpers but they have not been discussed till so we are using the example of motors.

Fix/UnFix Tag

The **Fix/UnFix Tag** tool is used to fix or unfix the tags assigned to the components. Note that if you fix a component tag then it will not get updated automatically when you re-tag the components in the drawing. The procedure to use this tool is given next.

- Click on the Fix/UnFix Tag tool from the Edit Components drop-down. The Fix/UnFix Component Tag dialog box will be displayed; refer to Figure-12.

Figure-12. Fix or UnFix Component Tag dialog box

- Select the desired radio button from the dialog box. For example, select the Force selected tags to Fixed radio button; then later all the selected tags will get fixed.
- Click on the **OK** button from the dialog box. You are asked to select the tags that you want to change for their tags fixed/unfixed status.
- Select the tags and press **ENTER**, the properties for selected tags will get modified as per the selected radio button.

COPY CATALOG ASSIGNMENT

The Copy Catalog Assignment tool is used to copy the meta data from an existing component and apply it to the other component. The procedure to use this tool is given next.

- Click on the **Copy Catalog Assignment** tool from the **Edit Components** drop-down. You are asked to select the master component.

- Click on the tag of the component from which you want to copy the meta data; refer to Figure-13. The **Copy Catalog Assignment** dialog box will be displayed; refer to Figure-14.

Figure-13. Component tag to be selected

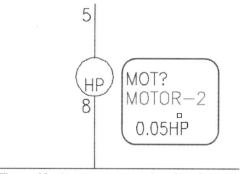

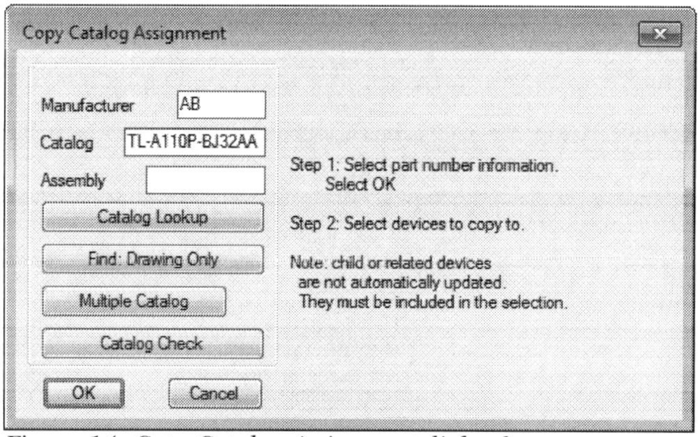

Figure-14. Copy Catalog Assignment dialog box

- Edit the details if you want to and then click on the OK button from the dialog box. You are asked to select the target component.
- Select the target component. Some of my friends might get the Update Related Components dialog box; refer to Figure-15. Click on the Yes-Update button if you want to update all similar components in the drawing or choose the Skip button to keep others as they are.
- Few of my friends might also get the Different symbol block

names dialog box; refer to Figure-16. This dialog box comes if you are assigning properties to an incompatible component. For example, if you are coping properties of a motor to a switch then this dialog box will come. If it is need of your drawing to assign such properties then click on the **OK** button and then **Overwrite** otherwise cancel it and take your decision.

Figure-15. Update Related Components dialog box

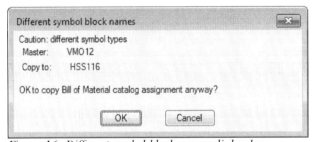
Figure-16. Different symbol block names dialog box

USER TABLE DATA

The **User Table Data** tool allows you to enter any data that you want to include with the components for their identity or purpose. The procedure to use this tool is given next.

- Click on the **User Table Data** tool from the **Edit Components** drop-down. The **Edit User Table Data** dialog box will be displayed; refer to Figure-17.

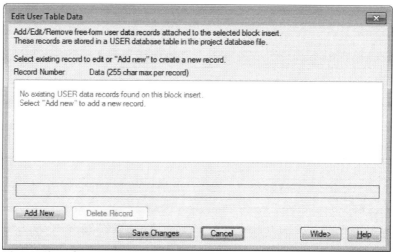
Figure-17. Edit User Table Data dialog box

- Click on the **Add New** button from the dialog box. The **Add New USER data record** dialog box will be displayed; refer to Figure-18.

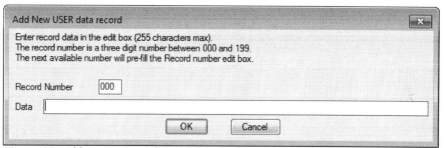
Figure-18. Add New USER data record dialog box

- Enter the desired data in the field and click on the OK button from the dialog box. The record will be added for the component; refer to Figure-19.

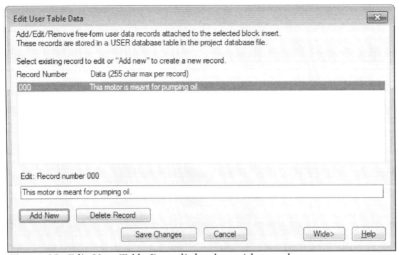

Figure-19. Edit User Table Data dialog box with record

- Click on the **Save Changes** button from the dialog box. The user record will be added to the component.

DELETE COMPONENT

As the name suggests, the **Delete Component** tool is used to delete the component. The procedure to do so is given next.

- Click on this tool from the **Edit Component** panel and click on the components that you want to delete.
- After selecting the components, Press **ENTER**. The components will be deleted and the **Search for/ Surf to Children** dialog box will be displayed; refer to Figure-20.

Figure-20. Search for or Surf to Children dialog box

- Click on the **OK** button if you want to search for the child components and want to delete them otherwise select the **NO** button from the dialog box.

COPY COMPONENT

The **Copy Component** tool is used to create copies of an existing component. The procedure to use this tool is given next.

- Click on the **Copy Component** tool from the **Edit Components** panel in the **Ribbon**. You are asked to select the component that you want to copy.
- Select the component. You are asked to specify the insertion points of the copied component.
- Click at the desired place in the drawing to create copy and then specify the desired parameters in the Insert/Edit Component dialog box.
- Click on the **OK** button from the dialog box after specifying the parameters. The copy will be created with the specified parameters.

EDIT CIRCUITS DROP-DOWN

The tools in the **Edit Circuits** drop-down are used to edit the circuits. There are three tools in this drop-down; refer to Figure-21. The procedure to use these tools is discussed next.

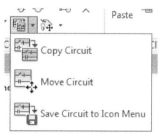

Figure-21. Edit Circuit drop-down

Copying Circuit

- Click on the **Copy Circuit** tool from the **Edit Circuits** drop-down. You are asked to select a circuit.
- Select the circuit and its components and then press ENTER. You are asked to specify the insertion point for the copied circuit.
- Click to specify the first point of insertion. You are asked to specify the second point of insertion; refer to Figure-22.

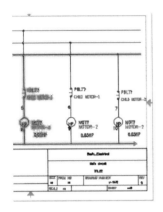
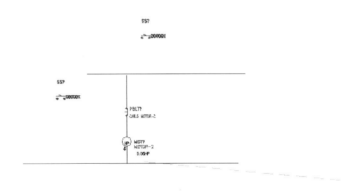

Figure-22. Copying a circuit

- Click to specify the insertion point. The circuit will be placed.
- Note that if you have gaps in the circuit then the **Gapped wire pointer problem** dialog box will be displayed; refer to Figure-23.

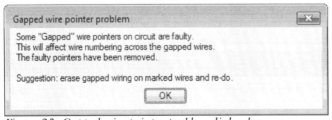

Figure-23. Gapped wire pointer problem dialog box

- Click on the **OK** button from the dialog box and modify the circuit as per your requirement.

Moving Circuit

- Click on the **Move Circuit** tool from the **Edit Circuits** drop-down. You are asked to select a circuit.
- Select the circuit and its components and then press ENTER. You are asked to specify the new base point for the circuit.
- Click to specify the base point. You are asked to specify the insertion point for the circuit. The circuit will be placed accordingly.

Saving Circuit to Icon Menu

- Click on the **Save Circuit to Icon Menu** tool from the **Edit Circuits** drop-down. The **Save Circuit to Icon Menu** dialog box will be displayed; refer to Figure-24.

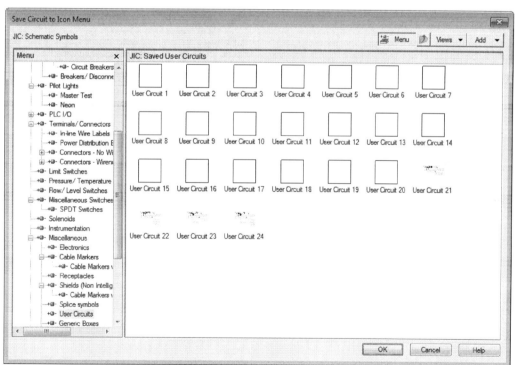

Figure-24. Save Circuit to Icon Menu dialog box

- Right-click on the desired slot from the dialog box. A shortcut menu will be displayed.
- Click on the **Properties** button from the shortcut menu. The **Properties-Circuit** dialog box will be displayed; refer to Figure-25.

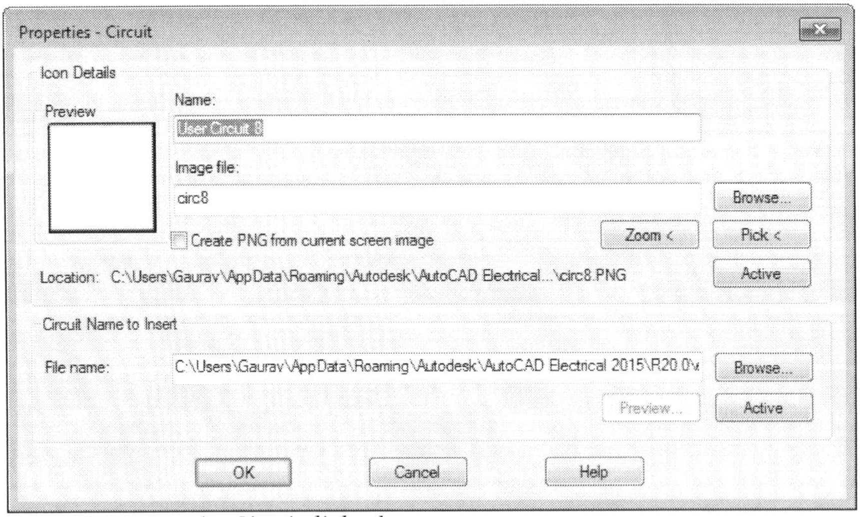

![Properties Circuit dialog box]

Figure-25. Properties Circuit dialog box

- Click on both the **Active** buttons one by one and click on the **OK** button from the dialog box. The current drawing will be saved as user defined circuit; refer to Figure-26.

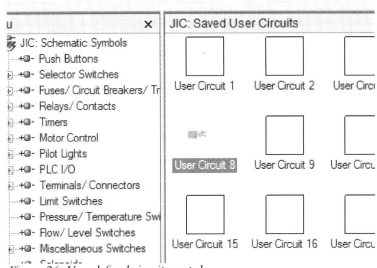

Figure-26. User defined circuit created

- Click on the **OK** button from the **Save Circuit to Icon Menu** dialog box.

TRANSFORMING COMPONENTS DROP-DOWN

The tools in this drop-down are used to transform the components. For example, you can change position of a component on wire, you can align components, and so on. Various tools in this drop-down are discussed next.

Scooting

Scooting is the movement of a component, tag, wire number etc. along the connected wire. The procedure to use this tool is given next.

- Click on the **Scoot** tool from the **Transform Component** drop-down. You are asked to select component, wire or wire number from the drawing area.
- Click on the desired entity. A box will be attached to the cursor representing the component; refer to Figure-27.

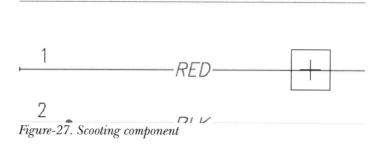
Figure-27. Scooting component

- Click at the desire point to specify the new location of the entity; refer to Figure-28.

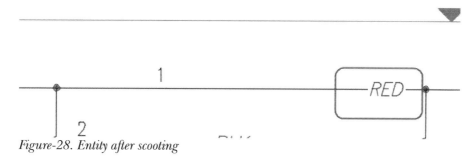
Figure-28. Entity after scooting

Aligning Components

- Click on the **Align** tool from the **Transform Component** drop-down. You are asked to select an entity as the horizontal/vertical reference for the other components.
- Select the component. The reference line will be created from the center of the selected component; refer to

Figure-29. Also, you are asked to selected the to be aligned components.

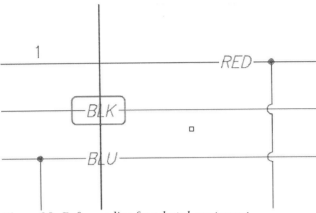

Figure-29. Reference line for selected component

- Click on the entities that you want to align and press **ENTER**. The entities will be aligned; refer to Figure-30.

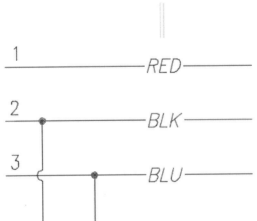

Figure-30. Entities after alignment

Moving Component

You can move components in a circuit without creating gaps in the circuit. The procedure is given next.

- Click on the **Move Component** tool from the **Transform Components** drop-down. You are asked to select a component.
- Click on the component. The component will get attached to the cursor; refer to Figure-31.

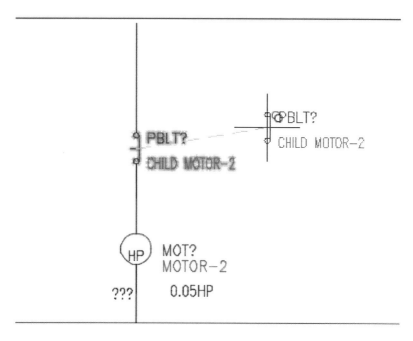

Figure-31. Component attached to the cursor

- Click at the new position to place the component. The component will be placed and the circuit will be modified to fill the gap; refer to Figure-32.

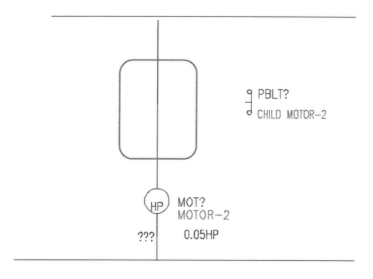

Figure-32. Circuit after moving component

Revering or Flipping Component

- Click on the **Reverse/Flip Component** tool from the **Transform Components** drop-down. The **Reverse/Flip Component** dialog box will be displayed as shown in Figure-33.

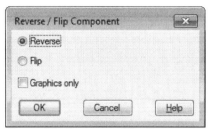

Figure-33. Reverse or Flip Component dialog box

- Select the desired radio button, whether you want to reverse the component or you want to flip the component.
- Click on the **OK** button from the dialog box. You are asked to select the component that you want to flip or reverse.
- Select the component, it will get flipped or reversed as per the selected option. Refer to Figure-34.

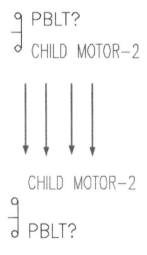

Figure-34. Reverse component

We will discuss about the other two tools in the **Transform Components** drop-down later in the book when we will be discussing PLCs.

RE-TAGGING COMPONENTS

You can assign a new tag to the components by replacing the older one using the **Retag Components** tool. The procedure to use this tool is given next.

- Click on the **Retag Components** tool from the **Retag Components** drop-down in the **Edit Components** panel in the **Ribbon**; refer to Figure-35. The **Retag Components** dialog box will be displayed; refer to Figure-36.

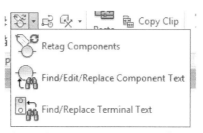

Figure-35. Retag Components drop-down

Figure-36. Retag Components dialog box

- Select the desired radio button from the dialog box and click on the **OK** button. The components will be re-tagged according to the selected radio button.

TOGGLE NO/NC

You can toggle between **Normally Open(NO)** and **Normally Closed(NC)** variants of components by using this tool. The procedure to use this tool is given next.

- Click on the **Toggle NO/NC** tool ![icon] from the **Edit Components** panel. You are asked to select the component for toggling NC/NO.
- Select the component, its status will be changed; refer to Figure-37.

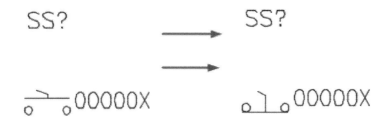

Figure-37. Component after toggling

SWAP/UPDATE BLOCK

This tool is used to swap or update the block of a component. The procedure to use this tool is given next.

Swapping

- Click on the **Swap/Update Block** tool from the **Edit Components** panel. The **Swap Block/Update Block/Library Swap** dialog box will be displayed; refer to Figure-38.
- Click on the desired radio button for **Swap a Block** area, the options related to swap will be activated in the dialog box.
- Select the desired option for source of swapping by selecting **Pick new block icon menu, Pick new block "just like"**, or **Browser to new block from file selection dialog** radio button.
- Click on the OK button from the dialog box. If you have selected the **Browser to new block from file selection dialog** radio button, then the dialog box will be displayed prompting you to select the blocks.
- Select the drawing of block. You are asked to select the component.
- Select the component and it will be swapped with the selected block.

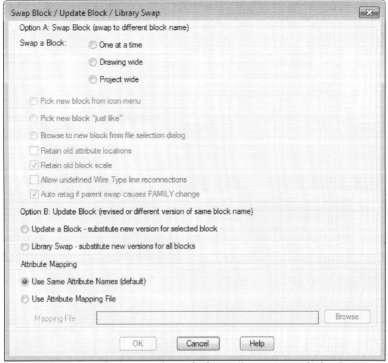

Figure-38. Swap Block or Update Block or Library Swap dialog box

Updating

- Click on the desired radio button from the Option B: Update area of the dialog box. If you select the **Update a Block** radio button then new block will be placed for selected component. If you select the **Library Swap** radio button then all the blocks will be updated.
- After selecting the desired radio button, click on the **OK** button from the dialog box.
- If you selected the **Update a Block** radio button then you will be prompted to select an example of block. Select the block of component(or component) from the drawing area and press **ENTER**. The **Update Block** dialog box will be displayed; refer to Figure-39. Click on the **Browse** button and select the desired component block. Click on the **Project** button if you want to update the block for all project drawings or click on the **Active Drawing** button to update the block in current drawing only.

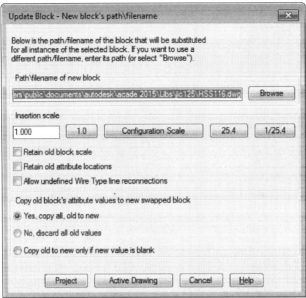
Figure-39. Update Block dialog box

- If you have selected Library Swap radio button in the **Swap Block/Update Block/Library Swap** dialog box then the Library Swap dialog box will be displayed; refer to Figure-40. Click on the **Browse** button and select the directory of component blocks. Click on the **Project** button if you want to update the block for all project drawings or click on the **Active Drawing** button to update the block in current drawing only.

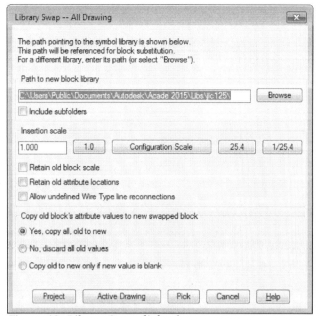

Figure-40. Library Swap dialog box

EDIT ATTRIBUTE DROP-DOWN

The options in the **Edit Attribute** drop-down are used to edit the attributes of the components; refer to Figure-41. These tools are common to all the AutoCAD platform software. An attribute is used to relate the blocks with the real word information. These tools perform the action as their names suggest. You can also perform all these tasks by using the Properties Manager. To use the Properties Manager, click on the component's attribute and enter **PROPERTIES** at the Command Prompt. The Properties Manager will be displayed with the relative options; refer to Figure-42. Click in the desired field and change the attributes as required.

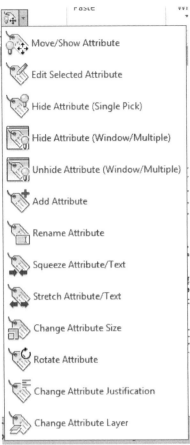

Figure-41. Edit Attribute drop-down

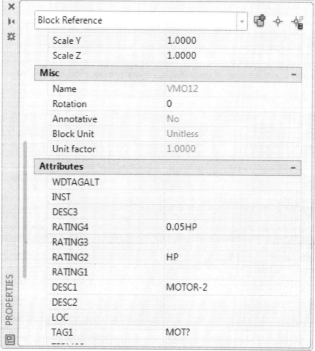

Figure-42. Properties Manager

CROSS REFERENCES DROP-DOWN

The tools in this drop-down are used to manage cross-references for the components; refer to Figure-43. The tools in this drop-down are discussed next.

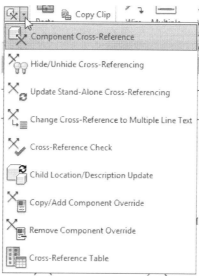

Figure-43. Cross References drop-down

Component Cross-Reference

This tool is used to update cross-reference text of a component. The procedure is given next.

- Click on the **Component Cross-Reference** tool from the **Cross References** drop-down. The Component Cross-Reference dialog box will be displayed; refer to Figure-44.

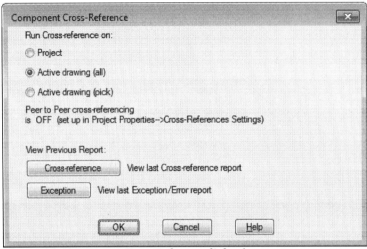

Figure-44. Component Cross-Reference dialog box

- Select the desired radio button from the dialog box and click on the **OK** button. The **Error/Exception Report** dialog box will be displayed; refer to Figure-45.

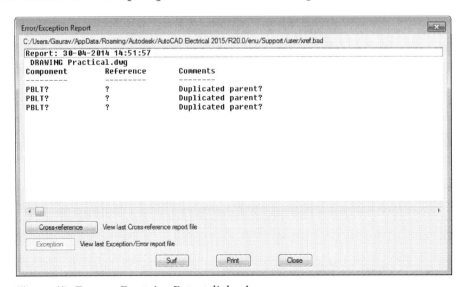

Figure-45. Error or Exception Report dialog box

- From the above result, we can find out that the three motors are duplicated and are not referenced to any other component or circuit.
- Click on the **Cross-reference** button from the dialog box. The **Cross-Reference Report** dialog box will be displayed; refer to Figure-46.

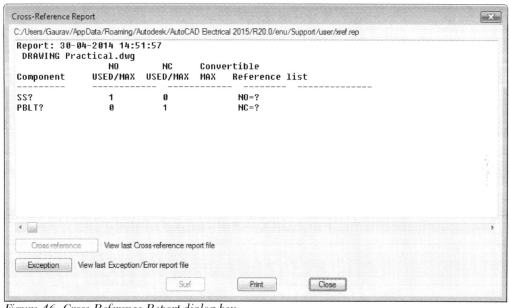

Figure-46. Cross Reference Report dialog box

- From the above result, we can find the components and their respective references in the drawing/project.
- After checking the results, click on the **Close** button to exit.

Hide/Unhide Cross-Referencing

The **Hide/Unhide Cross-Referencing** tool is used to hide or unhide the cross reference of a component in the drawing. Click on this tool from the **Cross References** drop-down and select the component of which you want to change the hide/unhide status.

Update Stand-Alone Cross-Referencing

The **Update Stand-Alone Cross-Referencing** tool is used to update the cross-references in the drawing or project. The procedure is given next.

- Click on **Update Stand-Alone Cross-Referencing** tool from the Cross References drop-down. The **Update Wire Signal and Stand-Alone Cross-Reference** dialog box will be displayed; refer to Figure-47.

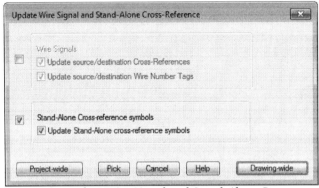

Figure-47. Update Wire Signal and Stand-Alone Cross-Reference dialog box

- Click on the **Drawing-wide** or **Project-wide** button to update all the cross-references in the drawing or project respectively. You can individually update the cross-references by selecting the Pick button from the dialog box.

In this same way, you can use the other tools in the **Cross References** drop-down. Rest of the tools in the **Edit Components** panel will be discussed in the chapter for PLCs.

Circuit Clipboard panel

The tools in this panel are used to manipulate circuits; refer to Figure-48. Click on the **Cut** or **Copy Clip** tool to copy any circuit in the drawing and using the **Paste** tool, you can paste the copied or cut circuits.

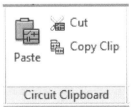

Figure-48. Circuit Clipboard panel

EDITING WIRES OR WIRE NUMBERS

The tools to edit wires/wire numbers are available in the **Edit Wires/Wire Numbers** panel in the **Ribbon**; refer to Figure-49. The tools in this panel are discussed next.

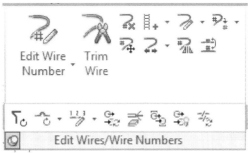

Figure-49. Edit Wires or Wire Numbers panel

Edit Wire Number

The Edit Wire Number tool is used to manually change the wire number. The procedure to use this tool is given next.

- Click on the down arrow below **Edit Wire Number** button in the **Ribbon**. The **Edit Wire Number** drop-down will be displayed; refer to Figure-50.

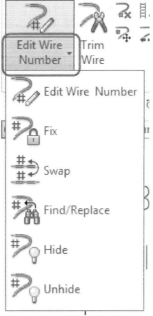

Figure-50. Edit Wire Number drop-down

- Click on the **Edit Wire Number** button from the drop-down. You are asked to select a wire or wire number.
- Select the wire or wire number that you want to change. The **Modify/Fix/UnFix** dialog box will be displayed; refer to Figure-51.

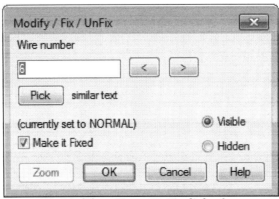

Figure-51. Modify or Fix or UnFix dialog box

- Click in the **Wire number** edit box and specify the desired value. You can use the **<** or **>** buttons to decrease or increase the wire number. Note that if the number is existing then you will be prompted to enter a different wire number.
- Clear the **Make it Fixed** check box, if you want to unfix the wire number.
- Select the **Hidden** radio button if you want to hide the wire number.
- Click on the **OK** button to apply the specified parameters.

Fix

The **Fix** tool is used to make a wire number fixed. To do so, click on the **Fix** tool from the **Edit Wire Number** drop-down. You are asked to select a wire number that you want to be fixed. Select the wire number and it will get fixed. Also, the color of the wire number will change to **CYAN** color.

Swap

The **Swap** tool is used to swap the wire numbers. Click on this tool from the **Edit Wire Number** drop-down and select two wire numbers one by one. The wire numbers will be swapped.

Find/Replace

The **Find/Replace** tool is available in the **Edit Wire Number** drop-down. This tool is used to find and replace the wire numbers in the drawing. The procedure to use this tool is given next.

- Click on the **Find/Replace** tool from the drop-down. The **Find/Replace Wire Numbers** dialog box will be displayed; refer to Figure-52.

Figure-52. Find or Replace Wire Numbers dialog box

- Click in the **Find** edit box and specify the wire number to be replaced.
- Click in the **Replace** edit box and specify the wire number with which it is to be replaced.
- Click on the **Go** button and then click on the **OK** button, the wire numbers will be replaced.

Hide and Unhide

The Hide tool is used to hide the wire numbers and Unhide tool is used to unhide the wire numbers. Click on the Hide tool and select the wire number that you want to hide. To unhide the wire number, click on the **Unhide** tool from the **Edit Wire Number** drop-down. You are asked to select a wire. Select the wire, the hidden wire number for that wire will be displayed.

Trim Wire

The **Trim Wire** tool is used to trim a wire connected between wire connections. The procedure to use this tool is given next.

- Click on the **Trim Wire** tool from the **Edit Wires/Wire Numbers** panel. You are asked to select the wire that is to be trimmed.
- Select the wire that you want to be trimmed; refer to Figure-53.

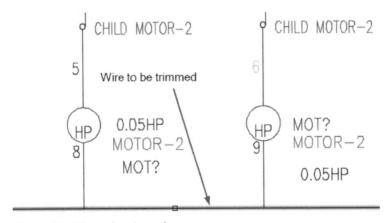

Figure-53. Wire to be trimmed

- On selecting the wire, it will be trimmed between the connecting ends; refer to Figure-54. Also, the wire numbering will be modified accordingly.

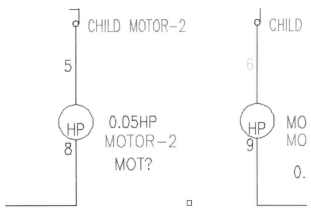
Figure-54. After trimming wire

Delete Wire Numbers

The **Delete Wire Numbers** tool is used to delete the wire numbers. Note that the remaining wire numbers are not modified on using this tool. To delete the wire numbers, click on this tool and then click on the wire numbers that you want to be deleted. The selected wire numbers will be deleted.

Move Wire Number

The **Move Wire Number** tool is used to move the wire numbers. To do so, click on **Move Wire Number** tool from the **Edit Wires/Wire Numbers** panel; you will be prompted to select a new location for the wire number. Click on the wire at the desired position. The wire number of that wire will move to the location you clicked on the wire.

Add Rung

The **Add Rung** tool is used to add a rung in any circuit or ladder. A rung is a wire connecting two parallel wiring lines. The procedure to use this tool is given next.

- Click on the **Add Rung** tool from the **Ladder Editing** drop-down; refer to Figure-55. You are asked to click at the desired location to add rung.
- Click between the two parallel wires. A rung will be created connecting both the wires; refer to Figure-56.

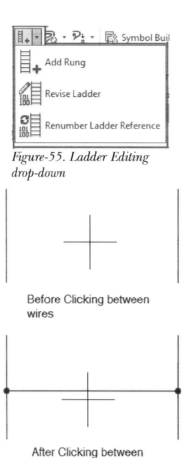

Figure-55. Ladder Editing drop-down

Figure-56. Rung connected

Revise Ladder

The **Revise Ladder** tool is available in the **Ladder Editing** drop-down. This tool is used to modify the ladder numbering. The procedure to use this tool is given next.

- Click on the **Revise Ladder** tool from the **Ladder Editing** drop-down. The **Modify Line Reference Numbers** dialog box will be displayed; refer to Figure-57.

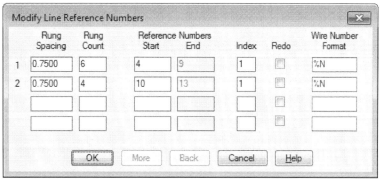

Figure-57. Modify Line Raferece Number dialog box

- Specify the desired parameters for the ladder numbering in the current drawing and click on the **OK** button from the dialog box. The ladder numbering will be modified accordingly.
- Note that to use this tool, you must have ladders in the drawing area.

Renumber Ladder Reference

The **Renumber Ladder Reference** tool is used to specify the reference point for the ladder renumbering. The procedure to use this tool is given next.

- Click on the **Renumber Ladder Reference** tool from the **Ladder Editing** drop-down. The **Renumber Ladders** dialog box will be displayed; refer to Figure-58.

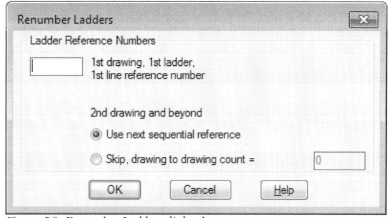

Figure-58. Renumber Ladders dialog box

- Click in the edit box and specify the desired reference number. Next, click on the **OK** button from the dialog box.

The **Select Drawings to Process** dialog box will be displayed; refer to Figure-59.

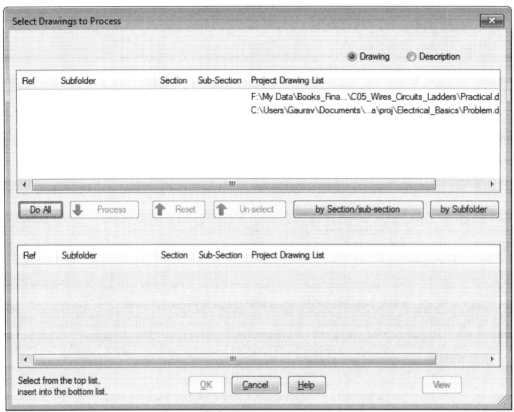
Figure-59. Select Drawings to Process dialog box

- Select the drawing and click on the **Process** button on which you want to apply the numbering. To select all click on the **Do All** button and then click on the **OK** button from the dialog box. The ladders will be renumbered as per the specified value.

Wire Editing

The tools in the **Wire Editing** drop-down are used to modify wires; refer to Figure-60. The tools in this drop-down are discussed next.

Figure-60. Wire Editing drop-down

Stretch Wire tool

The **Stretch Wire** tool is used to stretch the end point of a wire to connect it to the next intersecting wire. The procedure to use this tool is given next.

- Click on the **Stretch Wire** tool from the **Wire Editing** drop-down. You are asked to select the end point of the wire.
- Click on the wire near the end point. The wire will stretch till it intersect with the facing wire; refer to Figure-61.

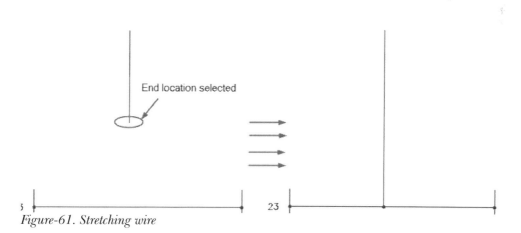

Figure-61. Stretching wire

Bend Wire tool

The **Bend Wire** tool is used to bend the wire at desired points. The procedure to use this tool is given next.

- Click on the **Bend Wire** tool from the **Wire Editing** drop-down. You are asked to select the point on first wire from which the bending will start.
- Click on the wire at desired point; refer to Figure-62.

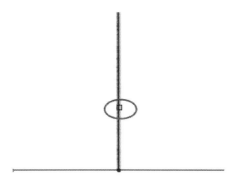

Figure-62. First point of bending

- Click on the second wire to specify the end point of bend. The wire bend will be created refer to Figure-63.

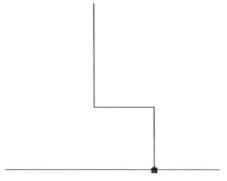

Figure-63. Wire bend after selecting second point

Show Wires

The **Show Wire** tool is used to display all the wire in the drawing. The procedure to use this tool is given next.

- Click on the **Show Wires** tool from the **Wire Editing** drop-down. The **Show Wires and Wire Number Pointers** dialog box will be displayed; refer to Figure-64.

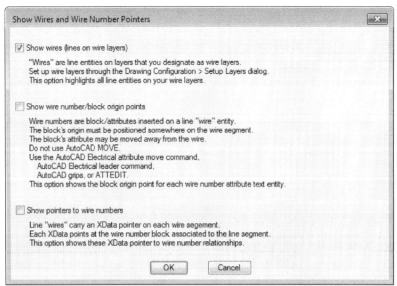

Figure-64. Show Wires and Wire Number Pointers dialog box

- Select the desired check boxes to highlight the respective entities in the drawing.
- Click on the **OK** button from the dialog box. The **Drawing Audit** dialog box will be displayed; refer to Figure-65.

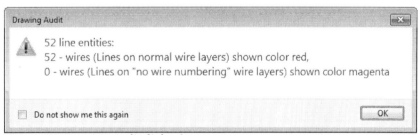

Figure-65. Drawing Audit dialog box

- Click on the **OK** button from the dialog box. The wires will be highlighted; refer to Figure-66.

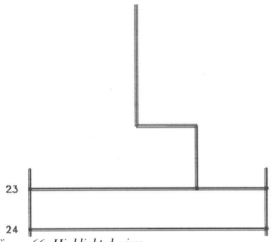
Figure-66. Highlighted wires

Check or Trace Wire tool

The **Check/Trace Wire** tool is used to highlight a wire in the drawing. The procedure to use this tool is given next.

- Click on the **Check/Trace Wire** tool from the **Wire Editing** drop-down. You are asked to select a wire segment.
- Click on the wire, its full length will be highlighted.
- Press the **SPACEBAR** to highlight the connected wire. Keep on pressing the **SPACEBAR** from keyboard and the next connected wire will be highlighted; refer to Figure-67.

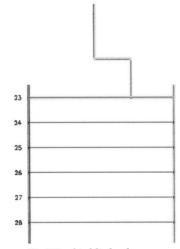
Figure-67. Wire highlighted

Wire Type Editing drop-down

The tools in the **Wire Type Editing** drop-down are used to edit the wire types; refer to Figure-68. These tools and their operations are discussed next.

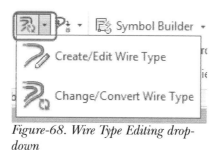

Figure-68. Wire Type Editing drop-down

Create/Edit Wire Type

The **Create/Edit Wire Type** tool is used to create or edit wire types. The procedure to use this tool is given next.

- Click on the **Create/Edit Wire Type** tool from the **Wire Type Editing** drop-down. The **Create/Edit Wire Type** dialog box will be displayed as shown in Figure-69.
- The fields in the table of this dialog box are used to specify parameters of common type of wires. If you want to modify the wire type then click on the desired field and specify the desired values. Note that **Layer Name** is automatically assigned on the basis of **Wire Color** and **Size** values.
- To add a new wire type, slide down the bottom of table and click in the **Wire Color** field; refer to Figure-70.

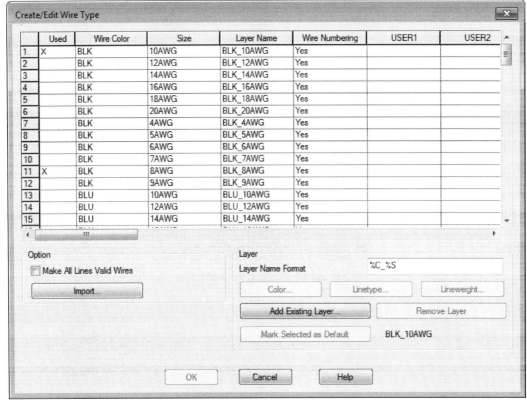

Figure-69. Create or Edit Wire Type dialog box

Figure-70. Wire Color field

- Specify the desired color. For example, **ORNG**.
- Click in the **Size** field next to it and specify the desired size. For example, **4AWG**.

- Click on the desired button in the **Layer** area and specify the related properties for the layers.
- After setting the desired parameters, click on the **OK** button from the dialog box. The wire type will be created.

Change/Convert Wire Type

The **Change/Convert Wire Type** tool is used to change the wire type of wires in the drawing. The procedure to use this tool is given next.

- Click on the **Change/Convert Wire Type** tool from the **Wire Type Editing** drop-down. The **Change/Convert Wire Type** dialog box will be displayed as shown in Figure-71.

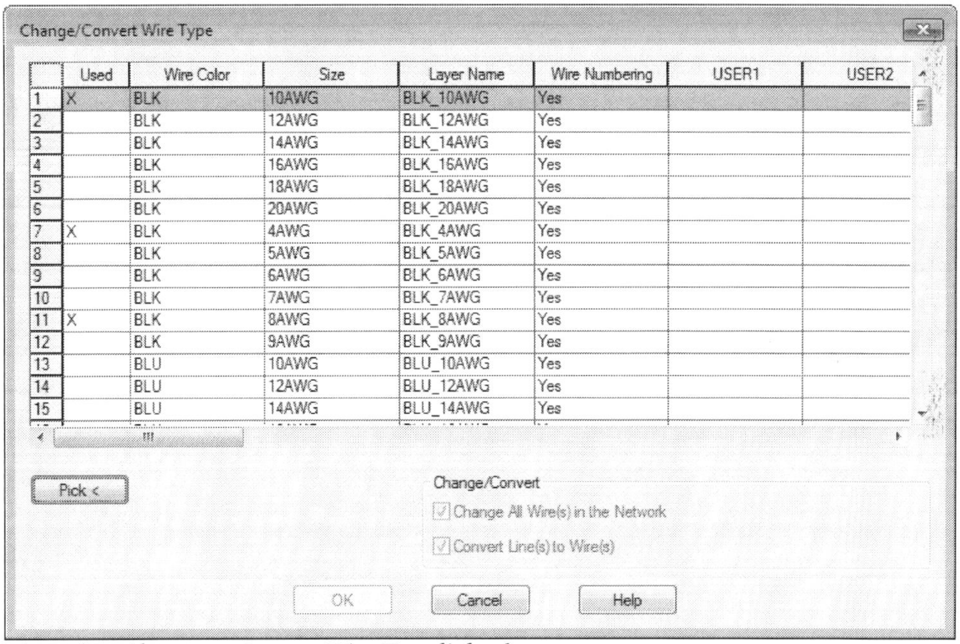

Figure-71. Change or Convert Wire Type dialog box

- Select the desired wire type from the list and click on the **OK** button from the dialog box. You are asked to select a wire.
- Select the wire from the drawing area. All the similar wires will be converted to the wire type selected in the dialog box.

Flip Wire Number

The **Flip Wire Number** tool is used to flip the wire number on the other side of the wire. The procedure to use this tool is given next.

- Click on the **Flip Wire Number** tool  from the **Edit Wires/Wire Numbers** panel in the **Ribbon**. You are asked to select a wire number.
- Select the wire number and it will be flipped to the other side of wire; refer to Figure-72.

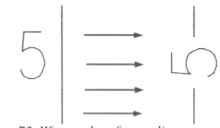

Figure-72. Wire number flipped

Toggle Wire Number In-Line

The **Toggle Wire Number In-Line** tools works in the same way as the **Flip Wire Number** tool works. After using this tool, the wire number gets inserted in the middle of the wire; refer to Figure-73.

Figure-73. Wire number after toggling

Till this point, we have discussed all the general tools used in editing components, circuits and wires. In the next chapter, we will discuss PLC and its components.

PAGE LEFT BLANK INTENTIONALLY FOR STUDENT NOTES

PAGE LEFT BLANK INTENTIONALLY FOR STUDENT NOTES

PLCs and Components

Chapter 7

The major topics covered in this chapter are:

- *Introduction*
- *Inserting Parametric PLCs*
- *Inserting PLCs (Full Unit)*
- *Inserting Connectors*
- *Inserting Terminals*

INTRODUCTION

PLC is a solid state/ computerized industrial computer that performs discrete or sequential logic in a factory environment. It was originally developed to replace mechanical relays, timers, counters. PLCs are used successfully to execute complicated control operations in a plant. Its purpose is to monitor crucial process parameters and adjust process operations accordingly. A sequence of instructions is programmed by the user to the PLC memory and when the program is executed, the controller operates a system to the correct operating specifications.

PLC consists of three main parts: CPU, memory and I/O units.

CPU is the brain of PLC. It reads the input values from inputs, runs the program existed in the program memory and writes the output values to the output register. Memory is used to store different types of information in the binary structure form. The memory range of S7-200 is composed of three main parts as program, parameter, and retentive data fields. I/O units provide communication between PLC control systems.

Application of PLCs in manufacturing process

In an industrial set up PLCs are used to automate manufacturing and assembly processes. By 'Process' we mean a step-by step procedure whereby a product is manufactured and assembled. It is the responsibility of the product engineering department to plan for manufacture of new or modified products. Other processes might involve the filling and capping of bottles, the printing of newspapers, or the assembly of automobiles etc. in many such manufacturing situations, PLC plays an important role in carrying out the various processes.

Examples of PLC Applications

Internal Combustion Engine Monitoring
A PLC acquires data recorded from sensors located at the internal combustion engine. Measurements taken include

water temperature, oil temperature, RPMs, torque, exhaust temperature, oil pressure, manifold pressure, and timing.

Carburetor Production Testing
PLCs provide on-line analysis of automotive carburetors in a production assembly line. The systems significantly reduce the test time, while providing greater yield and better quality carburetors. Pressure, vacuum, and fuel and air flow are some of the variables tested.

Monitoring Automotive Production Machines
The system monitors total parts, rejected parts, parts produced, machine cycle time, and machine efficiency. Statistical data is available to the operator anytime or after each shift.

Power Steering Valve Assembly and Testing
The PLC system controls a machine to ensure proper balance of the valves and to maximize left and right turning ratios.

CHEMICAL AND PETROCHEMICAL

Ammonia and Ethylene Processing
Programmable controllers monitor and control large compressors used during ammonia and ethylene manufacturing. The PLC monitors bearing temperatures, operation of clearance pockets, compressor speed, power consumption, vibration, discharge temperatures, pressure, and suction flow.

Dyes
PLCs monitor and control the dye processing used in the textile industry. They match and blend colors to predetermined values.

Chemical Batching
The PLC controls the batching ratio of two or more materials in a continuous process. The system determines the rate of discharge of each material and keeps inventory records. Several batch recipes can be logged and retreived automatically or on command from the operator.

Fan Control
PLCs control fans based on levels of toxic gases in a chemical production environment. This system effectively removes gases when a preset level of contamination is reached. The PLC controls the fan start/stop, cycling, and speeds, so that safety levels are maintained while energy consumption is minimized.

Gas Transmission and Distribution
Programmable controllers monitor and regulate pressures and flows of gas transmission and distribution systems. Data is gathered and measured in the field and transmitted to the PLC system.

Pipeline Pump Station Control
PLCs control mainline and booster pumps for crude oil distribution. They measure flow, suction, discharge, and tank low/high limits. Possible communication with SCADA (Supervisory Control and Data Acquisition) systems can provide total supervision of the pipeline.

Oil Fields
PLCs provide on-site gathering and processing of data pertinent to characteristics such as depth and density of drilling rigs. The PLC controls and monitors the total rig operation and alerts the operator of any possible malfunctions.

GLASS PROCESSING

Annealing Lehr Control
PLCs control used to remove the internal stress from glass products. The system controls the operation by following the annealing temperature curve during the reheating, annealing, straining, and rapid cooling processes through different heating and cooling zones. Improvements are made in the ratio of good glass to scrap, reduction in labor cost, and energy utilization.

Glass Batching
PLCs control the batch weighing system according to stored glass formulas. The system also controls the electromagnetic feeders for in feed to and outfeed from the weigh hoppers, manual shut-off gates, and other equipment.

Cullet Weighing
PLCs direct the cullet system by controlling the vibratory cullet feeder, weight-belt scale, and shuttle conveyor. All sequences of operation and inventory of quantities weighed are kept by the PLC for future use.

Batch Transport
PLCs control the batch transport system, including reversible belt conveyors, transfer conveyors to the cullet house, holding hoppers, shuttle conveyors, and magnetic separators. The controller takes action after the discharge from the mixer and transfers the mixed batch to the furnace shuttle, where it is discharged to the full length of the furnace feed hopper.

MANUFACTURING/MACHINING

Production Machines
The PLC controls and monitors automatic production machines at high efficiency rates. It also monitors piece-count production and machine status. Corrective action can be taken immediately if the PLC detects a failure.

Transfer Line Machines
PLCs monitor and control all transfer line machining station operations and the interlocking between each station. The system receives inputs from the operator to check the operating conditions on the line-mounted controls and reports any malfunctions. This arrangement provides greater machine efficiency, higher quality products, and lower scrap levels.

Wire Machine
The controller monitors the time and ON/OFF cycles of a wiredrawing machine. The system provides ramping control

and synchronization of electric motor drives. All cycles are recorded and reported on demand to obtain the machine's efficiency as calculated by the PLC.

Tool Changing
The PLC controls a synchronous metal cutting machine with several tool groups. The system keeps track of when each tool should be replaced, based on the number of parts it manufactures. It also displays the count and replacements of all the tool groups.

Paint Spraying
PLCs control the painting sequences in auto manufacturing. The operator or a host computer enters style and color information and tracks the part through the conveyor until it reaches the spray booth. The controller decodes the part information and then controls the spray guns to paint the part. The spray gun movement is optimized to conserve paint and increase part throughput.

MATERIALS HANDLING

Automatic Plating Line
The PLC controls a set pattern for the automated hoist, which can traverse left, right, up, and down through the various plating solutions. The system knows where the hoist is at all times.

Storage and Retrieval Systems
A PLC is used to load parts and carry them in totes in the storage and retrieval system. The controller tracks information like a lane numbers, the parts assigned to specific lanes, and the quantity of parts in a particular lane. This PLC arrangement allows rapid changes in the status of parts loaded or unloaded from the system. The controller also provides inventory printouts and informs the operator of any malfunctions.

Conveyor Systems
The system controls all of the sequential operations, alarms, and safety logic necessary to load and circulate parts on a

main line conveyor. It also sorts products to their correct lanes and can schedule lane sorting to optimize palletizer duty. Records detailing the ratio of good parts to rejects can be obtained at the end of each shift.

Automated Warehousing
The PLC controls and optimizes the movement of stacking cranes and provides high turnaround of materials requests in an automated, high-cube, vertical warehouse. The PLC also controls aisle conveyors and case palletizers to significantly reduce manpower requirements. Inventory control figures are maintained and can be provided on request.

METALS

Steel Making
The PLC controls and operates furnaces to produce metal in accordance with preset specifications. The controller also calculates oxygen requirements, alloy additions, and power requirements.

Loading and Unloading of Alloys
Through accurate weighing and loading sequences, the system controls and monitors the quantity of coal, iron ore, and limestone to be melted. It can also control the unloading sequence of the steel to a torpedo car.

Continuous Casting
PLCs direct the molten steel transport ladle to the continuous- casting machine, where the steel is poured into a water-cooled mold for solidification.

Cold Rolling
PLCs control the conversion of semifinished products into finished goods through cold-rolling mills. The system controls motor speed to obtain correct tension and provide adequate gauging of the rolled material.

Aluminum Making
Controllers monitor the refining process, in which impurities are removed from bauxite by heat and chemicals. The system grinds and mixes the ore with chemicals and then pumps them into pressure containers, where they are heated, filtered, and combined with more chemicals.

POWER

Plant Power System
The programmable controller regulates the proper distribution of available electricity, gas, or steam. In addition, the PLC monitors powerhouse facilities, schedules distribution of energy, and generates distribution reports. The PLC controls the loads during operation of the plant, as well as the automatic load shedding or restoring during power outages.

Energy Management
Through the reading of inside and outside temperatures, the PLC controls heating and cooling units in a manufacturing plant. The PLC system controls the loads, cycling them during predetermined cycles and keeping track of how long each should be on or off during the cycle time. The system provides scheduled reports on the amount of energy used by the heating and cooling units.

Coal Fluidization Processing
The controller monitors how much energy is generated from a given amount of coal and regulates the coal crushing and mixing with crushed limestone. The PLC monitors and controls burning rates, temperatures generated, sequencing of valves, and analog control of jet valves.

Compressor Efficiency Control
PLCs control several compressors at a typical compressor station. The system handles safety interlocks, startup/shutdown sequences, and compressor cycling. The PLCs keep compressors running at maximum efficiency using the nonlinear curves of the compressors.

PULP AND PAPER

Pulp Batch Blending
The PLC controls sequence operation, ingredient measurement, and recipe storage for the blending process. The system allows operators to modify batch entries of each quantity, if necessary, and provides hardcopy printouts for inventory control and for accounting of ingredients used.

Batch Preparation for Paper-Making Processing
Applications include control of the complete stock preparation system for paper manufacturing. Recipes for each batch tank are selected and adjusted via operator entries. PLCs can control feedback logic for chemical addition based on tank level measurement signals. At the completion of each shift, the PLC system provides management reports on materials use.

Paper Mill Digester
PLCs control the process of making paper pulp from wood chips. The system calculates and controls the amount of chips based on density and digester volume. Then, the percent of required cooking liquors is calculated and these amounts are added to the sequence. The PLC ramps and holds the cooking temperature until the cooking is completed.

Paper Mill Production
The controller regulates the average basis weight and moisture variable for paper grade. The system manipulates the steam flow valves, adjusts the stock valves to regulate weight, and monitors and controls total flow.

RUBBER AND PLASTIC

Tire-Curing Press Monitoring
The PLC performs individual press monitoring for time, pressure, and temperature during each press cycle. The system alerts the operator of any press malfunctions. Information concerning machine status is stored in tables for later use. Report generation printouts for each shift include a summary of good cures and press downtime due to malfunctions.

Tire Manufacturing
Programmable controllers are used for tire press/cure systems to control the sequencing of events that transforms a raw tire into a tire fit for the road. This control includes molding the tread pattern and curing the rubber to obtain road-resistant characteristics. This PLC application substantially reduces the space required and increases reliability of the system and the quality of the product.

Rubber Production
PLCs provide accurate scale control, mixer logic functions, and multiple formula operation of carbon black, oil, and pigment used in the production of rubber. The system maximizes utilization of machine tools during production schedules, tracks in-process inventories, and reduces time and personnel required to supervise the production activity and the shift-end reports.

Plastic Injection Molding
A PLC system controls variables, such as temperature and pressure, which are used to optimize the injection molding process. The system provides closed-loop injection, where several velocity levels can be programmed to maintain consistent filling, reduce surface defects, and shorten cycle time.

I know that the above list of applications is quite long but it is important to know the applications of PLC thoroughly before we start working on PLCs in AutoCAD Electrical.

Specifications of PLCs
In real-world, the PLCs are specified by the following parameters.

- Input Voltages
- Output Voltages
- Number of I/O terminals

Now, we are ready to work on the electrical drawings containing PLCs.

INSERTING PLCS (PARAMETRIC)

There are two tools in AutoCAD Electrical to insert PLCs; **PLC (Parametric)** and **PLC (Full Units)**. These tools are available in the **Insert PLCs** drop-down in the **Insert Components** panel; refer to Figure-1. The PLC (Parametric) tool allows you to insert the PLCs based on your inputs whereas the PLC (Full Units) allow you to insert the ready made PLC modules in the drawing. The procedure to use the PLC (Parametric) tool is given next.

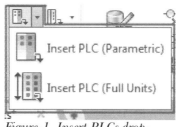

Figure-1. Insert PLCs drop-down

- Click on the **Insert PLC (Parametric)** tool from the **Insert PLCs** drop-down in the **Insert Components** panel. The **PLC Parametric Selection** dialog box will be displayed; refer to Figure-2.
- Click on the + signs to expand the categories and select the desired type for your PLC.
- Select the desired PLC from the table displayed at the bottom in the dialog box.
- Click on the desired radio button from the **Graphic Style** area displayed on the left in the dialog box. Note that the preview of displayed style is also displayed in the thumbnail.

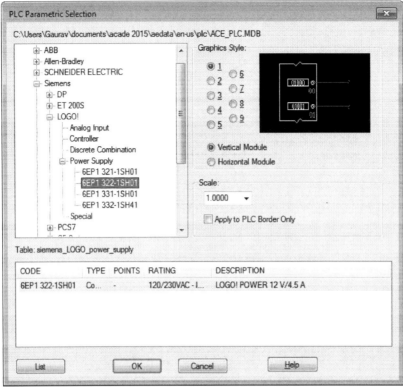

Figure-2. PLC Parametric Selection dialog box

- Click in the **Scale** edit box and specify the desired scale factor to make the PLC smaller or larger.
- Click on the **OK** button from the dialog box. The PLC will get attached to the cursor; refer to Figure-3.

Figure-3. PLC attached to the cursor

- Click in the drawing area to place the PLC. The **Module Layout** dialog box will be displayed; refer to Figure-4.

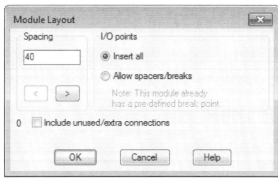

Figure-4. Module Layout dialog box

- Specify the desired spacing between the consecutive input and output points in the PLC by using the **Spacing** edit box in this dialog box.
- Select the **Allow spacers/breaks** radio button if you want to add spacers or break lines at the pre-defined break points in the PLC.
- If you want to display all the terminals whether they are used or unused then select the **Include unused/extra connections** check box.
- Click on the **OK** button from the dialog box. The **I/O Point** dialog box will be displayed; refer to Figure-5.

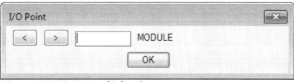

Figure-5. I/O Point dialog box

- Specify a number to the module by using the MODULE edit box of this dialog box and click on the **OK** button from the dialog box. The PLC will be placed.

INSERT PLC (FULL UNITS)

The **Insert PLC (Full Units)** tool is used to insert the full PLC units as per the requirement. Note that in this case, you cannot control the number of connections, display style and so on for the selected PLC. The procedure to use this tool is given next.

- Click on the **Insert PLC (Full Units)** tool from the **Insert PLCs** drop-down. The **Insert Component** dialog box will be displayed with the PLC components; refer to Figure-6.

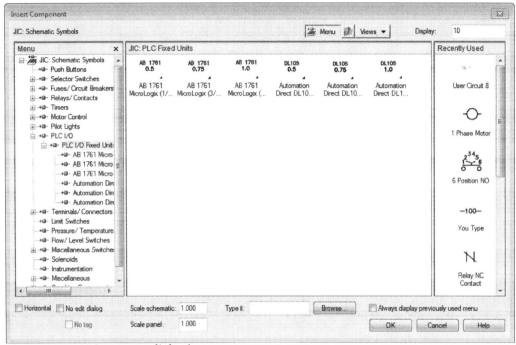

Figure-6. Insert Component dialog box

- Click on the desired category and select a plc of your requirement. The PLC will get attached to the cursor; refer to Figure-7.
- Click in the drawing area to place the PLC. The **Edit PLC Module** dialog box will be displayed; refer to Figure-8.

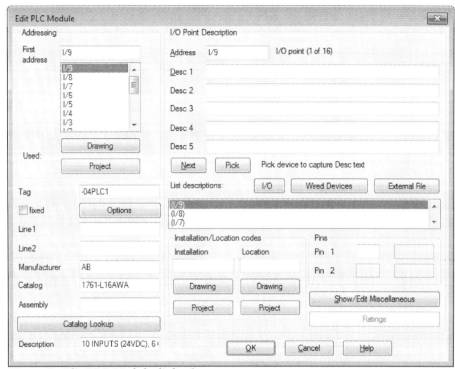
Figure-8. Edit PLC Module dialog box

- Specify the desired parameters in the dialog box and click on the **OK** button from the dialog box. The options in this dialog box are discussed next (its very important to understand the options in this dialog box).

Addressing Area

The options in this area are used to manage input or output addresses of the PLC.

First Address

Specifies the first I/O address for the PLC module. The addresses with I are meant for input like I/9, I/8, and so on. The addresses with O are meant for output like O/3, O/4, and so on. Select the desired address from the list to specify it as first I/O address.

Used Area

The buttons in this area are used to select I/O addresses used in earlier drawings or projects. The procedure to use these options is given next.

- Select the **Drawing** button from the area. The **PLC I/O Point List (this drawing)** dialog box will be displayed with the list of I/O points assigned already; refer to Figure-9.

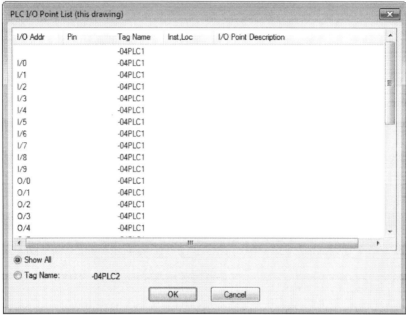

Figure-9. PLC IO Point List dialog box

- Select the desired tag name from the list to use it for the PLC.
- In the similar way, you can use **Project** button.

Note that editing the first address, I/O point address, or catalog information for a plc module that was imported using the Unity Pro Export to Spreadsheet tool may result in problems when you export the data back to Unity Pro. An alert displays to ask whether you want to proceed with the changes.

Tag

The Tag edit box is used to specify a unique identifier assigned to the module. The tag value can be manually typed in the edit box.

Options

The **Options** button is used to change the default format of tags for PLCs. On selecting this button, the **Option: Tag Format "Family" Override** dialog box is displayed; refer to Figure-10. Note that you can insert a fixed text string for the %F part of the tag format. Retag Component can then use this override format value to calculate a new tag for the PLC module. For example, a certain PLC module must always have an "IO" family tag value instead of "PLC" so that retag, for example, assigns IO-100 instead of PLC100. To achieve this tag override you would enter "IO-%N" for the tag override format.

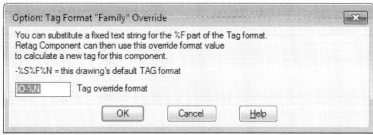

Figure-10. Option dialog box

Line1/Line2

These edit boxes are used to specify optional description text for the module. May be used to identify the relative location of the module in the I/O assembly (example: Rack # and Slot #).

Manufacturer

The **Manufacturer** edit box is used to list the manufacturer number for the module. Enter a value or select one by using the **Catalog Lookup** tool.

Catalog

Lists the catalog number for the module. Enter a value or select one from the **Catalog Lookup**.

Assembly

Lists the assembly code for the module. The Assembly code is used to link multiple part numbers together.

Catalog Lookup

The **Catalog Lookup** tool is used to display the data in catalog related to the selected component. Click on the **Catalog Lookup** tool and the **Catalog Information** dialog box will be displayed; refer to Figure-11. The options of this dialog box have already been discussed.

Figure-11. Catalog Information dialog box

Description

Specify the optional line of description text in the **Description** edit box. May be used to identify the module type (for example, "16 Discrete Inputs - 24VDC")

I/O Point Description Area
Address

This edit box is used to specify the I/O address assignment.

Description 1-5

These edit boxes are used to specify the optional description text. Enter up to five lines of description attribute text.

Next/Pick
The Next button is used to select the next I/O Point for specifying the description. Pick button is used to select the desired I/O point directly from the drawing.

List descriptions
The options in this area are used to list the I/O point descriptions currently assigned to each I/O point on the module or connected, wired devices in a pick list.

Pins
The edit boxes in this area are used to assign pin numbers to the pins that are physically located on the module.

Show/Edit Miscellaneous
The **Show/Edit Miscellaneous** button is used to view or edit any attributes that are not predefined AutoCAD Electrical attributes.

Ratings
This edit box is used to specify values for each ratings attribute. You can enter up to 12 ratings attributes on a module. Select **Defaults** to display a list of default values. Note: If Ratings is unavailable, the module you are editing does not carry rating attributes.

CONNECTORS
Connectors are the parts that are used to connect wirings from sources to equipment. The tools to add or modify the connectors are available in the **Insert Connectors** drop-down in the **Insert Components** panel; refer to Figure-12. Various tools in this drop-down are discussed next.

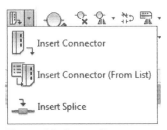

Figure-12. Insert Connector drop-down

Insert Connector

The **Insert Connector** tool is used to insert connectors in the drawing. The procedure to use this tool is given next.

- Click on the **Insert Connector** tool from the **Insert Connectors** drop-down. The **Insert Connector** dialog box will be displayed; refer to Figure-13.

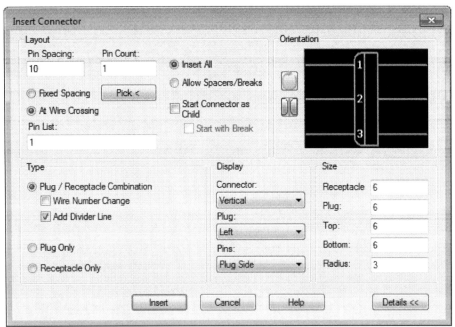

Figure-13. Insert Connector dialog box

- Click in the **Pin Spacing** edit box and specify the space between two consecutive pins of the connector.
- Click in the **Pin Count** edit box and specify the number of pins. Alternatively, click on the **Pick<** button and draw a construction line of desired length. The number of pins

will be automatically assigned based on the length of line and space between two pins.
- Select the **At Wire Crossing** radio button if you want to set the pins of connector at the crossing of wires with the connector.
- Click on the **Details>>** button to expand the dialog box.
- From the **Type** area, select the desired radio button to change the type of connector. After selecting the radio button from this area, you can check the preview in the thumbnail in this dialog box.
- If you want the wire numbers to be changed after the insertion of the connectors, then select the **Wire Number Change** check box.
- Select the desired options from the **Display** area to change the display style of the connector.
- The edit boxes in the **Size** area are used to change the size of connector.
- After specifying the desired parameters, click on the **Insert** button from the dialog box. The connector will get attached to the cursor.
- Click in the drawing area to place the connector. If you click on the ladder's wire then the pins of connector will automatically adjust according to the ladder; refer to Figure-14.

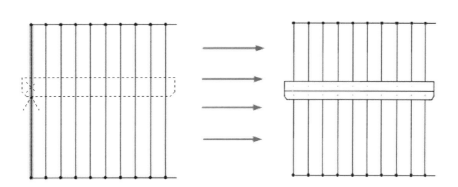

Figure-14. Placing connectors on a ladder

Insert Connector (From List)

The **Insert Connector (From List)** tool is used to insert the connectors from an XML file created by Inventor. In Inventor, you can create model of the connector and then you can export it into XML format. The procedure to use this tool is given next.

- Click on the **Insert Connector (From List)** tool from the **Insert Connectors** drop-down. The **Autodesk Inventor Professional Import File Selection** dialog box will be displayed; refer to Figure-15.

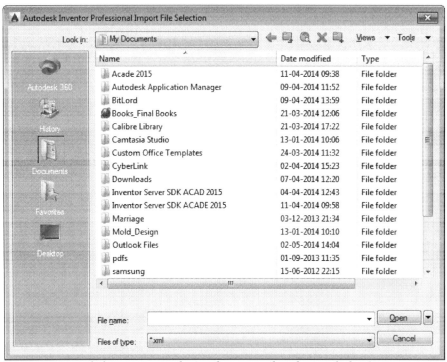

Figure-15. Autodesk Inventor Professional Import File Selection dialog box

- Select the desired file and click on the **Open** button. List of connectors will be displayed.
- Select the desired connector and place it in the drawing.

Insert Splice

The **Insert Splice** tool is used to insert splice in the wiring of connection points of connectors. Click on this tool to insert splice, the **Insert Component** dialog box will be displayed. The procedure to use this dialog box have already been discussed.

TERMINALS

Terminals are used to connect wires to the devices. In real world, terminals are used to connect wires so that they do not cause sparking at high load. You can use the **Insert Component** dialog box to insert terminals in the drawing.

The tools to edit terminals are available in the expanded **Edit Components** panel. Various tools related to terminals are discussed next.

Inserting Terminals from Catalog Browser

- Click on the **Catalog Browser** tool from the **Insert Components** panel in the **Ribbon**. The Catalog Browser will be displayed.
- Select the **TRMS** category, or any category with terminals.
- Enter a search criteria and click on the Search button.
- Click a row in the results pane.

Do one of the following:
- Click one of the symbols associated to the catalog value.
- Click Browse to locate the symbol on disk. The symbol selected is automatically associated to that catalog value for future insertions.

You can double-click the row in the results pane to insert the default or only symbol associated to the catalog value.

- Specify the insertion point in the drawing.

The symbol orientation matches the underlying wire. If there is no underlying wire, the selected orientation is inserted. The wire breaks automatically if the symbol lands on it.

Associate Terminals on the Same Drawing

You can use the **Associate Terminals** tool to associate two or more terminal symbols together. Associating schematic terminals combines the terminals into a single terminal block property definition. The number of schematic terminals that can be combined is limited to the number of levels defined for the block properties.

Associating a panel terminal provides a way to define a particular panel footprint to represent a schematic block property definition. The procedure to use this tool is given next.

- Click **Schematic** tab > **Edit Components** panel > **Associate Terminals** tool.
- Select a terminal symbol to use as the master. It is used as the basis for any terminal property definition.

Note that your terminal symbol must have block properties defined. To define block properties, right-click on the symbol and select Edit Component. In the Insert/Edit Terminal Symbol dialog box, click Block Properties.

- Select additional terminal symbols to add to the association of the master terminal.
- Press Enter to associate the terminals.
- The catalog data, block properties, Tag strip value, Installation code and Location code are copied from the master terminal and added to the terminals in the association.

Note: If the number of selected terminals exceeds the total number of levels defined in the block properties, an alert displays and the extra terminals are not added to the association.

You can use the other tools related to terminals in the same way.

Practical and Practice

Chapter 8

The major topics covered in this chapter are:

- **Introduction**
- **Practical 1**
- **Practical 2**
- **Practical 3**
- **Practice 1**
- **Practice 2**
- **Practice 3**
- **Practice 4**
- **Practice 5**

INTRODUCTION

In this chapter, we will create various electrical circuits and at the end of chapter a complete collection of electrical drawings is given for practice.

PRACTICAL 1

Create a schematic for 3 phase motor starter as shown in Figure-1.

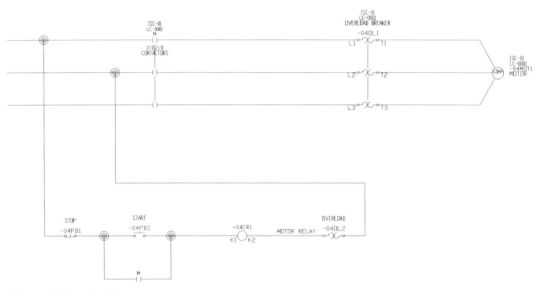

Figure-1. Practical 1

Starting AutoCAD Electrical drawing

- Click on the AutoCAD Electrical icon from the desktop or start AutoCAD Electrical by using the **Start** menu; refer to Figure-2.

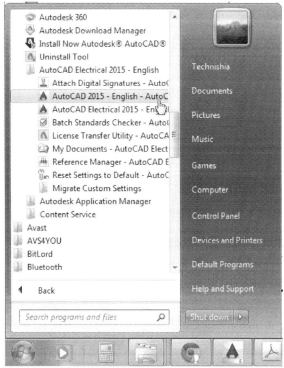

Figure-2. Starting AutoCAD using Start menu

- AutoCAD Electrical will open.
- Click on the down arrow below **Start Drawing** button from the template area. The list of drawing templates will be displayed; refer to Figure-3.

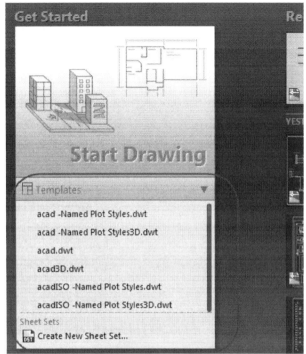
Figure-3. Drawing templates

- Select the **ACAD ELECTRICAL IEC.dwt** template from the list. A new drawing will be created with the selected template.
- Click on the buttons displayed in Figure-4 to hide the grid and off the grid snap.

Figure-4. Grid display and snap

Inserting wires

- Click on the **Multiple Bus** tool from the **Insert Wires/Wire Numbers** panel in the **Ribbon**. The **Multiple Wire Bus** dialog box will be displayed as shown in Figure-5.

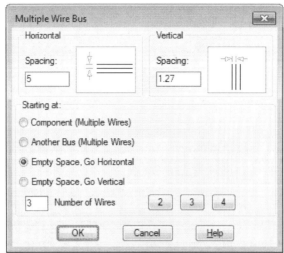

Figure-5. Multiple Wire Bus dialog box

- Click in the **Spacing** edit box in **Horizontal** area of the dialog box and specify the spacing as **5** in the edit box.
- Select the **Empty Space, Go Horizontal** radio button and specify the number of wires as 3 in the **Number of Wires** edit box.
- Click on the **OK** button from the dialog box. You are asked to specify the start point of the wires.
- Click in the drawing area at desired position to start wiring. You are asked to specify the end point of the wires.
- Click in the drawing area to specify the end point; refer to Figure-6.

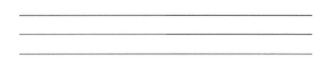

Figure-6. Wire bus created

- Click in the **Wire** tool from the **Wire** drop-down in the **Insert Wires/Wire Numbers** panel in the **Ribbon**. You are asked to specify the start point of the wire.
- Click on the top wire to specify the start point of the wire. You are asked to specify the next point of the wire.
- Click to specify the next point. Similarly, create all the wires of the circuit; refer to Figure-7.

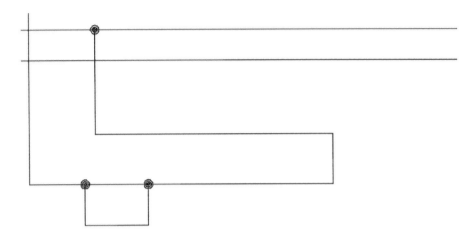

Figure-7. Complete wiring for current circuit

Inserting Components

- Click on the **Icon Menu** button from the **Insert Components** drop-down in the **Insert Components** panel in the **Ribbon**. The **Insert Component** dialog box will be displayed.
- Click on the **Motor Control** category from the dialog box. The components related to motor control will be displayed; refer to Figure-8.

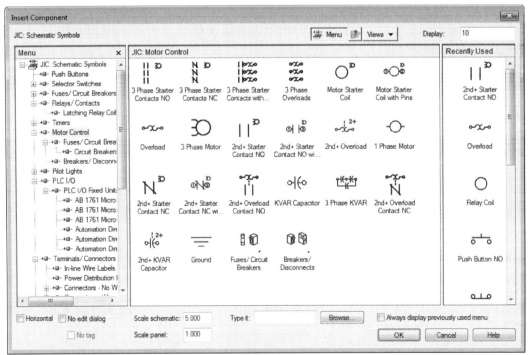

Figure-8. Insert Component dialog box

- Click on the **3 Phase Motor** component from the dialog box. The component will get attached to the cursor.
- Click at the end point of the center line of the wire bus; refer to Figure-9. The motor will be placed and the **Insert/Edit Component** dialog box will be displayed; refer to Figure-10.

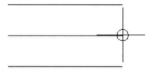

Figure-9. motor placement

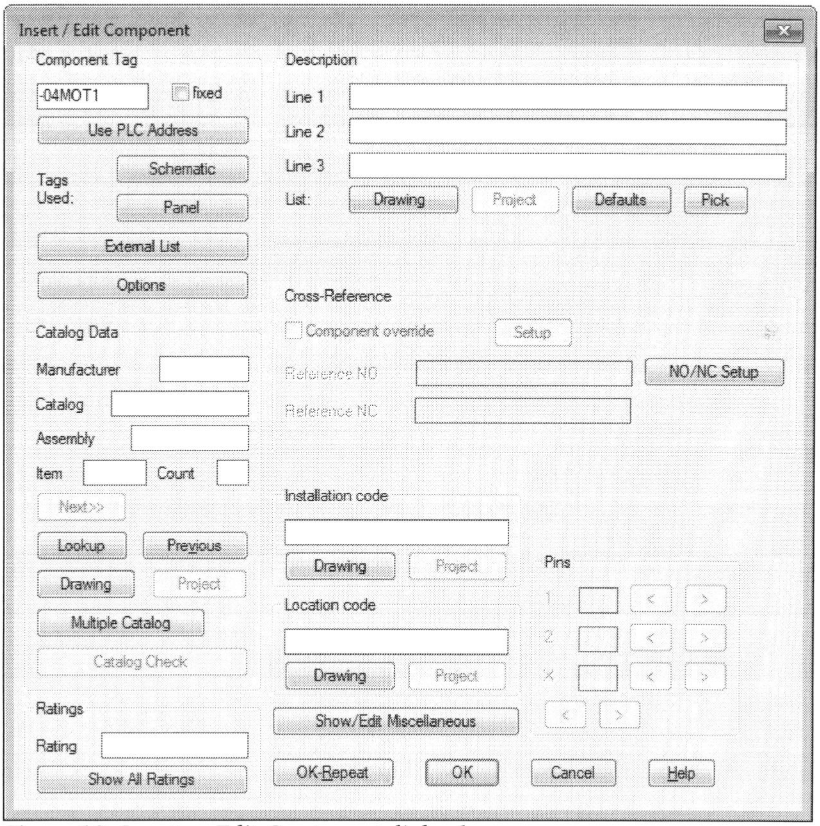

Figure-10. Insert or Edit Component dialog box

- Click on the **Lookup** tool from the **Catalog Data** area of the dialog box. The **Catalog Browser** will be displayed.

- Select the **CM111-FC1F518GSKCA** field under the **CATALOG** column from the **Catalog Browser**; refer to Figure-11. Note that this is a 3 Phase 1.5 HP motor with 1800 RPM speed.

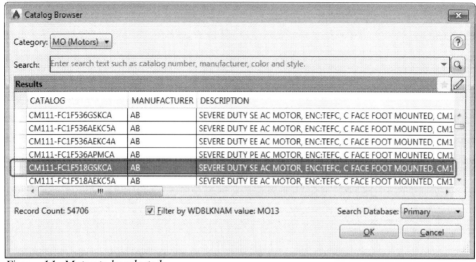

Figure-11. Motor to be selected

- Click on the **OK** button from the **Catalog Browser**.
- The catalog data will be updated automatically in the **Catalog Data** area of the **Insert/Edit Component** dialog box.
- Click on the **Show All Ratings** button from the **Ratings** area of the dialog box. The **View/Edit Rating Values** dialog box will be displayed; refer to Figure-12.
- Click in the **Rating 1** edit box and specify the value as **1.5HP** and remove the **Rating 4** edit box.
- Click on the **OK** button from the dialog box. The value will be displayed in the **Rating** edit box.

Figure-12. View or Edit Rating Values dialog box

- Click in the **Line 1** edit box in the **Description** area of the dialog box and specify the value as **MOTOR**.
- Click in the **Installation Code** edit box and specify the value as **ISC-01**.
- Click in the **Location Code** edit box and specify the value as **LC-0001**.
- Click on the **OK** button from the dialog box. If the **Assign Symbol To Catalog Number** dialog box is displayed (refer to Figure-13) then click on the **Map symbol to catalog number** button from it.

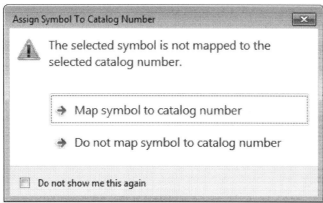

Figure-13. Assign Symbol To Catalog Number dialog box

- The motor after placement will be displayed as shown in Figure-14.

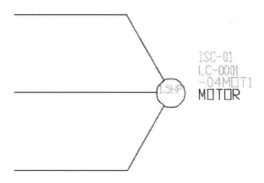

Figure-14. Motor after placement

Inserting 3 Phase Overload Circuit Breaker

- Click on the **Icon Menu** button from the **Insert Components** drop-down. The **Insert Component** dialog box will be displayed.
- Click on the **Motor Control** category and select the **3 Phase Overloads** icon from the dialog box; refer to Figure-15. The icon will get attached to the cursor; refer to Figure-16.

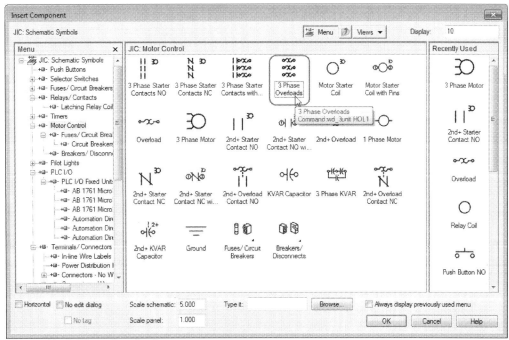

Figure-15. 3 Phase Overloads component

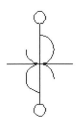

Figure-16. Icon attached to the cursor

- Click on the top wire of the wire bus; refer to Figure-17. The **Build Up or Down?** dialog box will be displayed; refer to Figure-18.

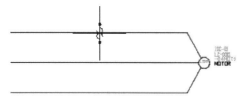

Figure-17. Placement of overload circuit breaker

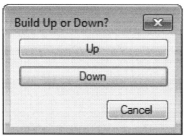

Figure-18. Build Up or Down dialog box

- Click on the **Down** button from the dialog box. The **Insert/ Edit Component** dialog box will be displayed.
- Click on the **Lookup** button from the **Catalog Data** area. The **Catalog Browser** dialog box will be displayed.
- Click on the **193-A1A1** field in the **CATALOG** column from the dialog box; refer to Figure-19.

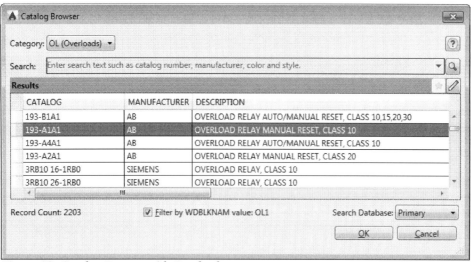

Figure-19. Catalog Browser with Overloads components

- Click on the **OK** button from the **Catalog Browser**. The details will appear in the **Catalog Data** area of the dialog box.
- Click in the **Line 1** edit box in the **Description** area of the dialog box. Specify the value as **OVERLOAD BREAKER**.
- For the **Installation Code** and **Location Code,** click on the **Drawing** button and select the earlier specified code.
- Click on the **OK** button from the dialog box. If the **Assign Symbol To Catalog Number** dialog box is displayed then click on the **Map symbol to catalog number** button. The breaker will be placed in the circuit; refer to Figure-20.

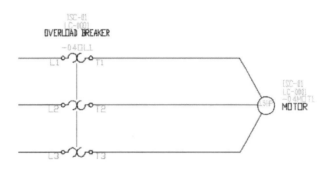

Figure-20. Overload Breaker

Inserting the Starter Contact

- Click on the **Icon Menu** button and select the **Motor Control** category.
- Click on the **3 Phase Starter Contacts NO** icon from the **Insert Component** dialog box. You are asked to specify the location of the contacts.
- Click on the top wire of the bus as did before; refer to Figure-21. The **Build Up or Down?** dialog box will be displayed.
- Click on the **Down** button from the dialog box. The contacts will be placed and the **Insert/Edit Child Component** dialog box will be displayed; refer to Figure-22.
- Click in the **Line 1** edit box in the **Description** area of the dialog box and specify the value as **CONTACTORS**.

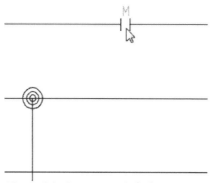

Figure-21. Contact symbol placement

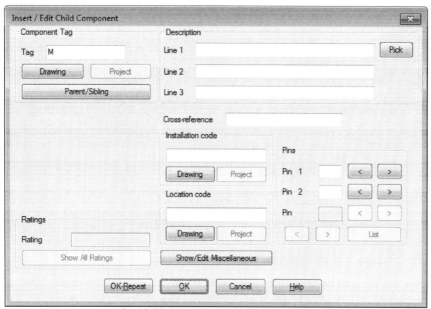

Figure-22. Insert or Edit Child Component dialog box

- For **Installation** and **Location** codes, click on the **Drawing** buttons and select the earlier specified values.
- Click on the **OK** button from the dialog box.

Similarly, insert the other components with the required descriptions. The schematic drawing after inserting all the components will be displayed as shown in Figure-23.

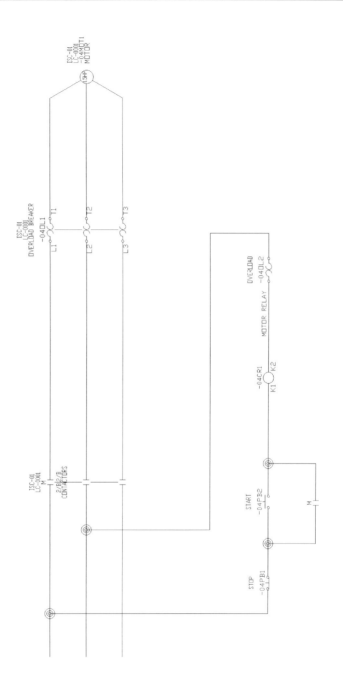

Figure-23. practical 1

Practical 2

In this practical, we will create a schematic drawing of circuit as shown in Figure-24. This circuit mainly emphasize on connector.

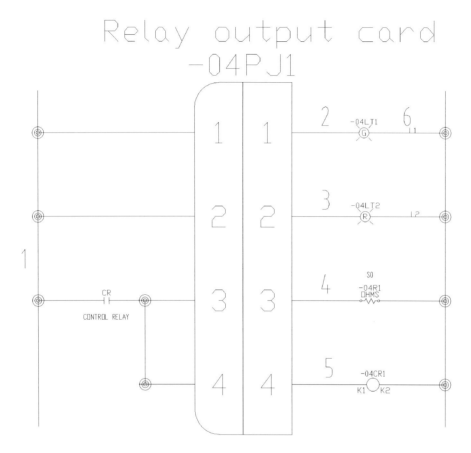

Figure-24. Practica 2

Starting AutoCAD Electrical drawing

- Click on the AutoCAD Electrical icon from the desktop or start AutoCAD Electrical by using the **Start** menu.

- AutoCAD Electrical will open.
- Click on the down arrow below **Start Drawing** button from the template area. The list of drawing templates will be displayed; refer to Figure-25.

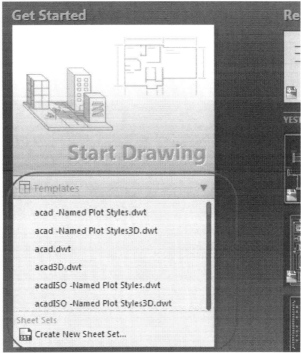

Figure-25. Drawing templates

- Select the **ACAD ELECTRICAL IEC.dwt** template from the list. A new drawing will be created with the selected template.
- Click on the buttons displayed in Figure-26 to hide the grid and off the grid snap.

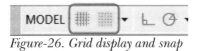

Figure-26. Grid display and snap

Inserting ladder

- Click on the **Insert Ladder** tool from the **Insert Ladders** drop-down in the **Insert Wires/Wire Numbers** panel. The **Insert Ladder** dialog box will be displayed; refer to Figure-27.

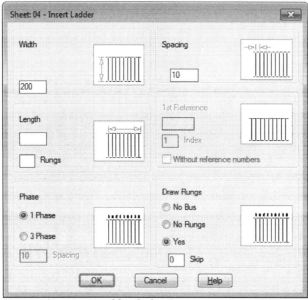

Figure-27. Insert Ladder dialog box

- Specify the **Width** value and **Spacing** value as **50** and **10**, respectively.
- Click in the **Rungs** edit box and specify the value as **4**.
- Click on the **OK** button from the dialog box. You are asked to specify the insertion point for the ladder.
- Click in the drawing area to place the ladder at the desired position. If your ladder is placed with horizontal orientation as shown in Figure-28, then rotate the complete ladder by 90 degree using the **Rotate** tool from the **Modify** panel in the **Home** tab of the **Ribbon**.

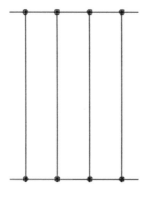

Figure-28. Ladder placed horizontally oriented

- After rotating the ladder, it will be displayed as shown in Figure-29.

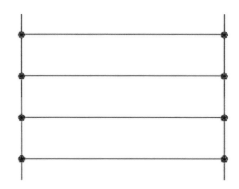

Figure-29. Rotated ladder

Inserting the Connector

- Click on the **Schematic** tab if you still in the **Home** tab; refer to Figure-30.

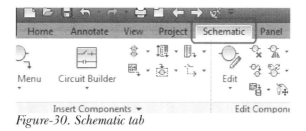

Figure-30. Schematic tab

- Click on the **Insert Connector** tool from the **Insert Connectors** drop-down in the **Insert Components** panel. The **Insert Connector** dialog box will be displayed as shown in Figure-31.

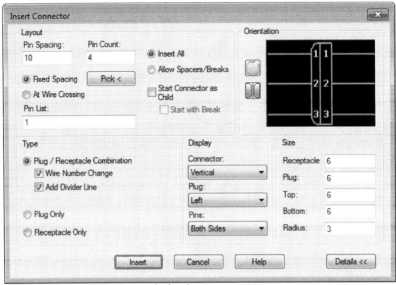

Figure-31. Insert Connector dialog box

- Click in the **Pin Spacing** edit box and specify the value as **10** (this is also the spacing between two consecutive rungs in the ladder earlier created).
- Click in the **Pin Count** edit box and specify the value as **4** (this is also the number of rungs in the ladder earlier created).
- Click in the **Connector** drop-down in the **Display** area of the dialog box and select the **Vertical** option from the list displayed if not selected.
- Click on **Insert** button from the dialog box. The connector will get attached to the cursor.
- Click on the top rung of the ladder as shown in Figure-32 to place the connector.

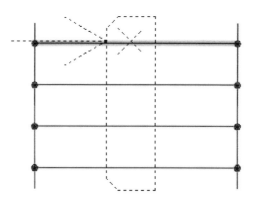

Figure-32. Placing the connector

- Select the connector and right click on it. A shortcut menu will be displayed as shown in Figure-33.

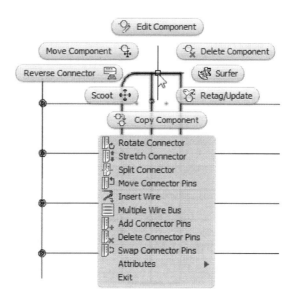

Figure-33. Menu for connector

- Move the cursor on the **Attributes** option in the menu. A cascading menu will be displayed; refer to Figure-34.

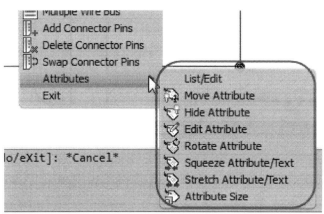

Figure-34. Cascading menu for attributes

- Click on the **Attribute Size** option from the cascading menu. The **Change Attribute Size** dialog box will be displayed; refer to Figure-35.

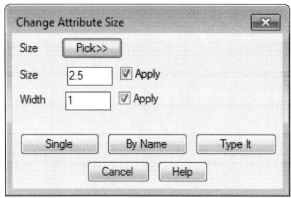

Figure-35. Change Attribute Size dialog box

- Make sure the size is **2.5** and width is **1**. Click on the **Single** button from the dialog box. You are asked to select the annotation objects.
- Click on the connector numbers and tags displayed in very small font to increase their size. Refer to Figure-36.

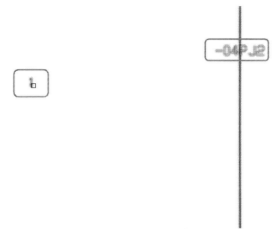

Figure-36. Numbers and tags in small font

- After increasing the size of annotations, the drawing will be displayed as shown in Figure-37.

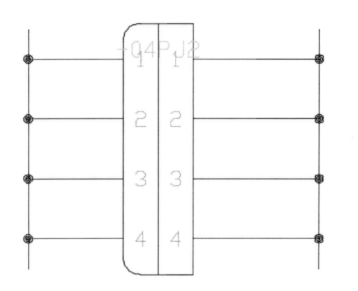

Figure-37. After increasing size of annotations

- Move the overlapping annotations to their proper places.

Connecting Wires

- Click on the **Wire** tool from the **Wires** drop-down in the **Insert Wires/Wire Numbers** panel. You are asked to specify the starting point of the wire.

- Click on the wire connected to connector number **3** in the left and connect it to the wire connected to connector number **4**; refer to Figure-38.

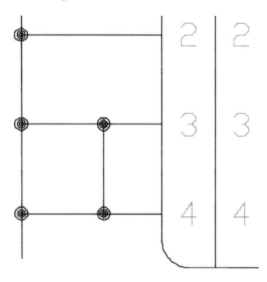

Figure-38. Wire connected

- Click on the **Trim Wire** tool from the **Edit Wires/Wire Numbers** panel in the **Ribbon**. You are asked to select the wire to be trimmed.
- Click on the wire as shown in Figure-39. Press ENTER to exit the **Trim Wire** tool.

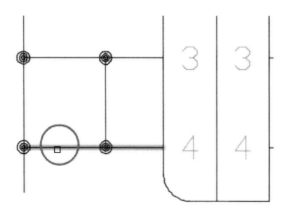

Figure-39. Wire to be trimmed

Inserting Components

- Click on the **Icon Menu** from the **Insert Components** drop-down in the **Insert Components** panel. The **Insert Component** dialog box will be displayed.
- Click on the **Relays/Contacts** category from the dialog box. The components related to relays and contacts will be displayed; refer to Figure-40.

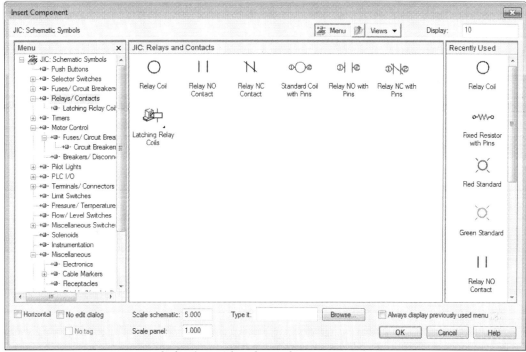

Figure-40. Insert Component dialog box with Relay and Contacts

- Click on the **Relay NO Contact** component from the dialog box. The component will get attached to the cursor.
- Place the component on the wire connected to Connector **3**; refer to Figure-41. The **Insert/Edit Child Component** dialog box will be displayed; refer to Figure-42.

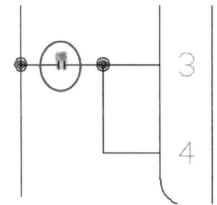

Figure-41. Relay placed

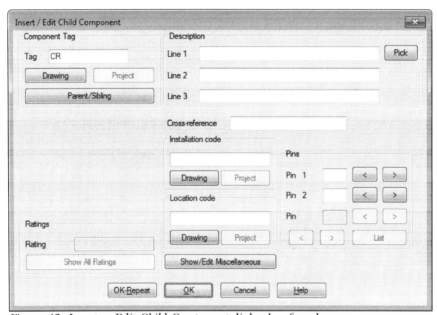

Figure-42. Insert or Edit Child Component dialog box for relay

- Click in the **Line 1** edit box and specify the value as **Control Relay**.
- Click on the **OK** button from the dialog box.
- Similarly, place the other components in the circuit. The schematic after placing all the components will be displayed as shown in Figure-43.

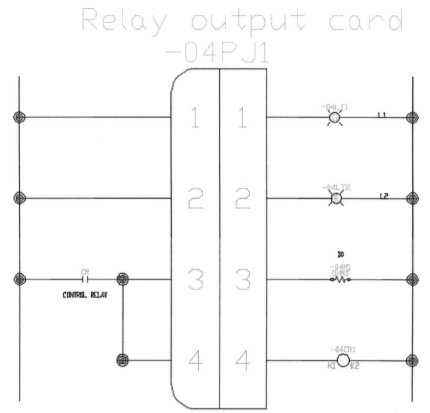

Figure-43. Schematic after placing the components

Assigning Wire Numbers

- Click on the **Wire Numbers** button from the **Wire Numbers** drop-down in the **Insert Wires/Wire Numbers** panel. The **Wire Tagging** dialog box will be displayed; refer to Figure-44.

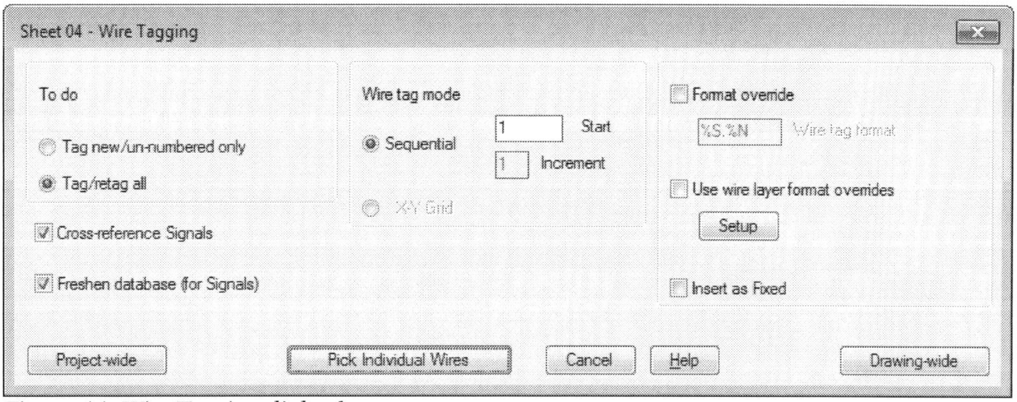

Figure-44. Wire Tagging dialog box

- Select the **Format override** check box. The edit box below it will become active.
- Click in the edit box and remove **%S.** from the edit box so that **%N** is left in the box; refer to Figure-45.

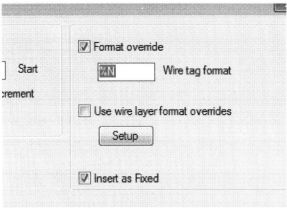

Figure-45. Format override

- Click on the Drawing-Wide button from the dialog box to assign the wire numbers to all the wires in the current drawing. The schematic after assigning the wire numbers will be displayed as shown in Figure-46.

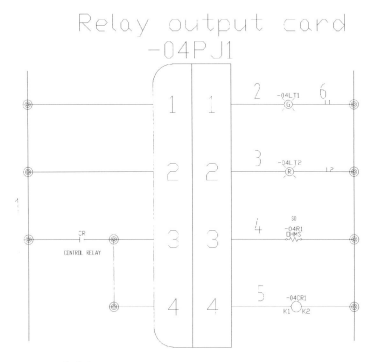

Figure-46. Practical 2

Practical 3
Create a PLC circuit as shown in Figure-47.

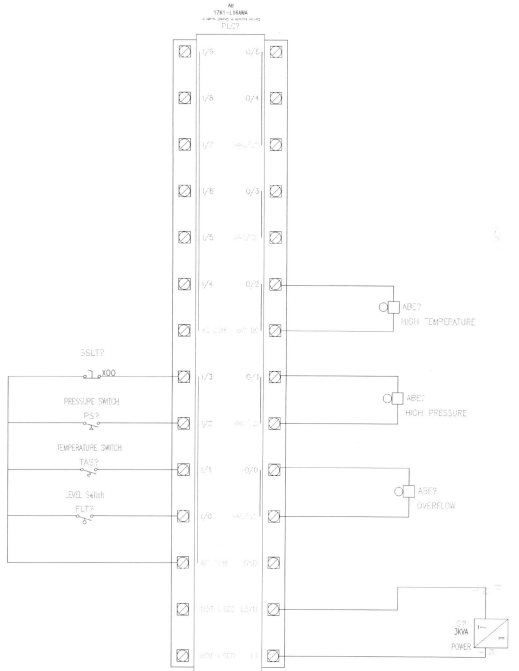

Figure-47. Practical 3

Starting AutoCAD Electrical drawing

- Click on the AutoCAD Electrical icon from the desktop or start AutoCAD Electrical by using the **Start** menu.
- AutoCAD Electrical will open.
- Click on the down arrow below **Start Drawing** button from the template area. The list of drawing templates will be displayed; refer to Figure-48.

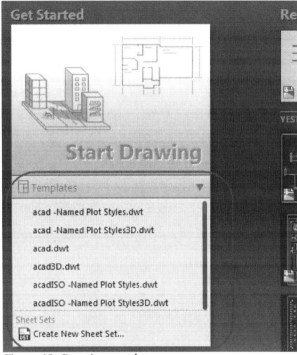

Figure-48. Drawing templates

- Select the **ACAD ELECTRICAL IEC.dwt** template from the list. A new drawing will be created with the selected template.
- Click on the buttons displayed in Figure-49 to hide the grid and off the grid snap.

Figure-49. Grid display and snap

Inserting PLC (Full Units)

- In this tutorial, we are going to use Allen Bradley's AB 1761-L16AWA model PLC. We recommend you to know the working of this PLC by browsing through the Allen Bradley's website.
- Click on the **Insert PLC (Full Units)** tool from the **Insert PLCs** drop-down in the **Insert Components** panel. The **Insert Component** dialog box will be displayed; refer to Figure-50.

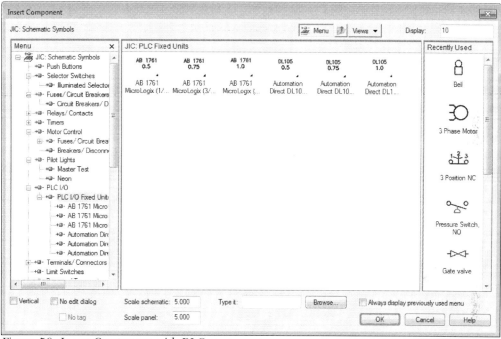

Figure-50. Insert Component with PLC components

- Click on the **AB 1761 Micrologix (1" Spacing)** category from the dialog box. The list of **PLCs** will be displayed; refer to Figure-51.
- Click on the **L16-AWA 10in/6out AC-DC/115AC-DC** component from the list. You are asked to specify the insertion point for the PLC.
- Click in the drawing area to place the PLC. The **Edit PLC Module** dialog box will be displayed; refer to Figure-52.

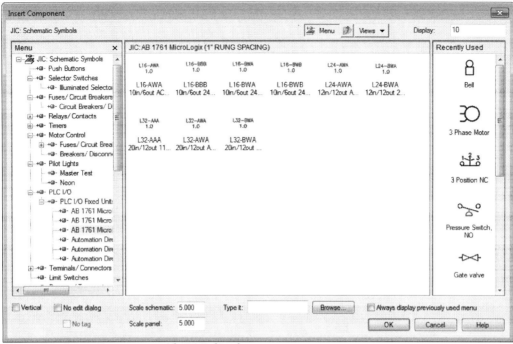

Figure-51. Insert Component with PLCs listed

Figure-52. Edit PLC Module dialog box

- You can specify descriptions for each input/output address as desired.
- Click on the **OK** button from the dialog box. The PLC will be placed with the specified descriptions; refer to Figure-53.

Figure-53. PLC placed in drawing

Adding Multiple Wire Bus

- Click on the **Multiple Bus** tool from the **Insert Wires/Wire Numbers** panel in the **Ribbon**. The Multiple Wire Bus dialog box will be displayed; refer to Figure-54.

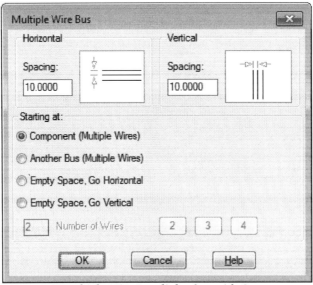

Figure-54. Multiple Wire Bus dialog box with Component radio button selected

- Make sure that the **Component (Multiple Wires)** radio button is selected in the dialog box.
- Click on the **OK** button from the dialog box. You are asked to make a window selection to select the ports of the component for connecting wires.
- Select the Input ports and AC COM as shown in Figure-55.
- Press ENTER from the keyboard and move the cursor. End point of the wire bus will get attached to the cursor; refer to Figure-56.
- Click at the desired distance to create the wire bus.

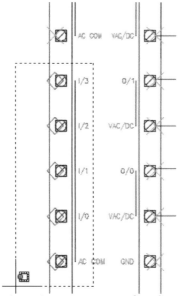

Figure-55. Input ports selected

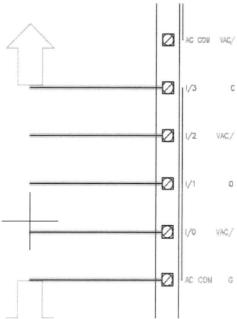

Figure-56. End point of wire bus attached to cursor

- Click on the **Wire** tool from the **Wires** drop-down in the **Insert Wires/Wire Numbers** panel. You are asked to specify the start point of the wire.

- Connect all the end points of the wire bus by creating a wire; refer to Figure-57.

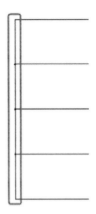

Figure-57. Wire connecting end points

Connecting Components

- Click on the **Icon Menu** button from the **Insert Components** drop-down in the **Insert Components** panel. The **Insert Component** dialog box will be displayed.
- Click on the **Selector Switches** category in the dialog box. The components in the dialog box will be displayed as shown in Figure-58. Note that you might need to change the scale of the component to properly display it in the drawing by using the edit boxes highlighted in a box in Figure-58.
- Select the 3 Position NC component. The component will get attached to the cursor.
- Click on the top wire of wire bus connected to I/3 address of PLC. The **Insert/Edit Component** dialog box will be displayed.
- Specify the desired parameters and click on the **OK** button from the dialog box.

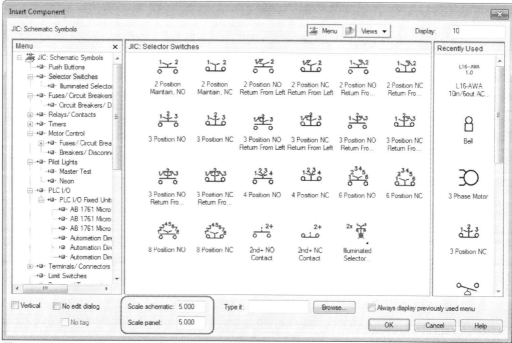

Figure-58. Insert Component dialog box with Selector Switches

- Similarly, you can add the other components in the wire bus. After connecting all the components in the wire bus, the schematic will be displayed as shown in Figure-59.

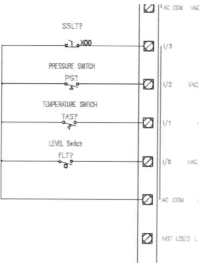

Figure-59. Wirebus after connecting components

Connecting Output Devices

- The right side of PLC is meant for output and power supply. Click on the **Wire** tool and build the circuit as shown in Figure-60.

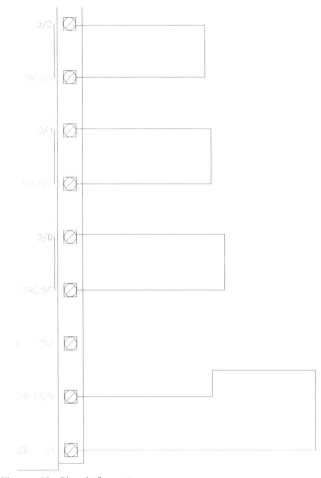

Figure-60. Circuit for output

- Using the **Icon Menu** tool, connect all the components to the circuit; refer to Figure-61.

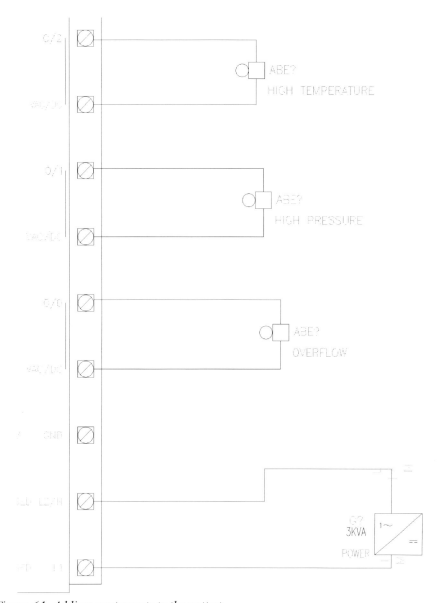

Figure-61. Adding components to the output

- After connecting all the components, the complete circuit will be displayed as shown in Figure-62.

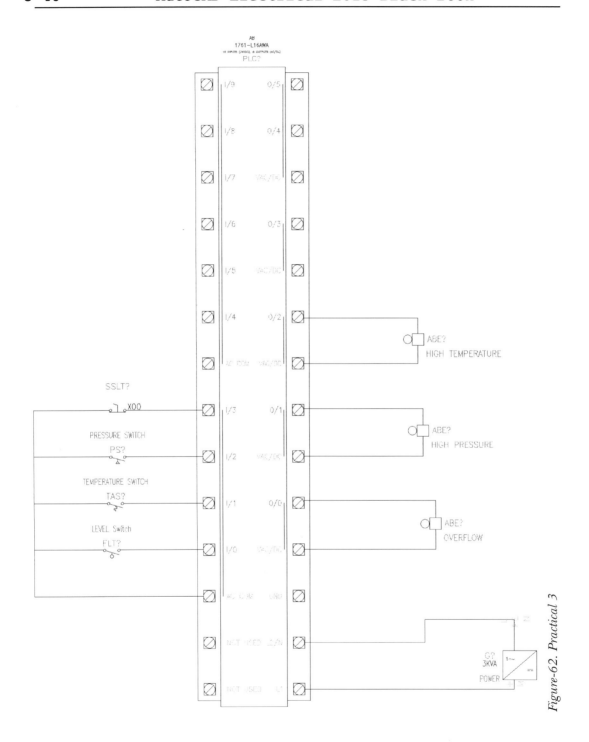

Figure-62. Practical 3

PRACTICE 1

Create the schematic drawing as shown in Figure-63.

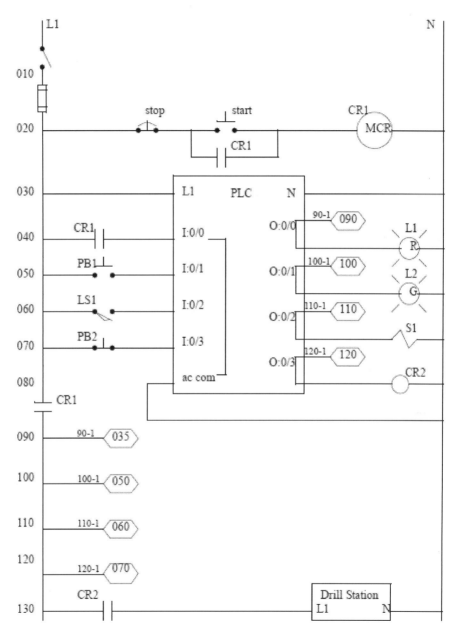

Figure-63. Practice 1

PRACTICE 2

Create the schematic drawing as shown in Figure-64.

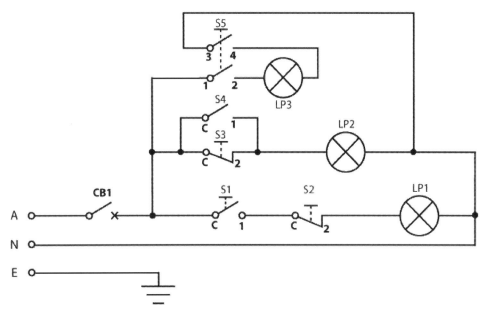

Figure-64. Practice 2

PRACTICE 3

Create the schematic drawing as shown in Figure-65.

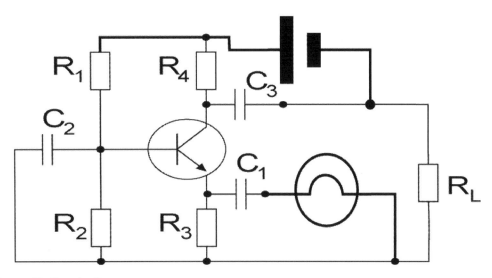

Figure-65. Practice 3

Practice 4

Create the schematic drawing as shown in Figure-66.

Output = $\overline{A}BC + A\overline{B}C + AB\overline{C} + ABC$

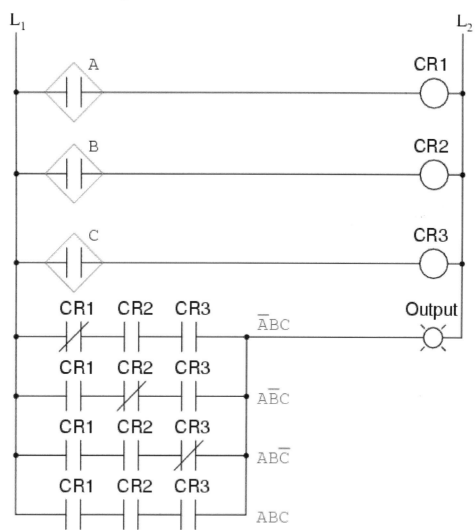

Figure-66. Practice 4

Practice 5

Create the schematic drawing as shown in Figure-67.

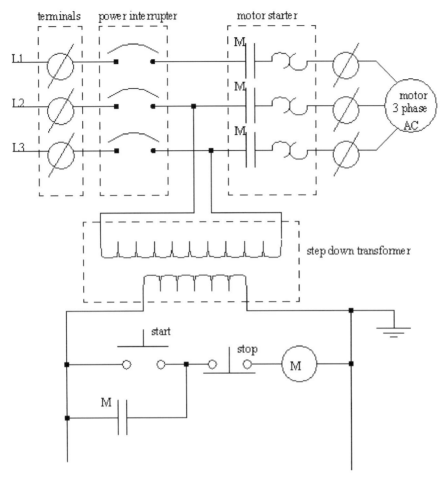

Figure-67. Practice 5

Panel Layout

Chapter 9

Topics Covered

The major topics covered in this chapter are:

- *Introduction*
- *Icon Menu for panels*
- *Schematic List*
- *Manual*
- *Manufacturers Menu*
- *Balloon*
- *Wire Annotation*
- *Panel Assembly*
- *Editor*
- *Table Generator*
- *Terminals*
- *Editing*

INTRODUCTION

In the previous chapters, you have learnt to create schematic circuit diagrams. After creating those circuit diagrams, the next step is to create panels. A panel is the box consisting of various electrical switches and PLCs to control the working of equipment. Refer to Figure-1. Note that the panel shown in the figure is back side panel of a machine. This panel is generally hidden from the operator. What an operator see is different type of panel; refer to Figure-2. We call this panel as User Control panel. In both the cases, the approach of designing is almost same but the interaction with the user is different. The User Control Panel is meant for Users so it can have push buttons, screen, sensors, key board and so on. On the other side, the back panel will be having relays, circuit breakers, sensors, connectors, plcs, switches and so on.

Figure-1. Panel

Figure-2. User Control panel

If we start linking the schematic drawings with the panel drawings then the common platform is the component tag and the location code. Suppose we have created a push button in the schematic with tag -04PB2 then in the panel layout you should insert the same push button with the same tag. The location of the Push button will be decided on the Location code. The components that are having same location code should be placed at the same place in the panel.

In AutoCAD Electrical, there is a separate tab for the tools related to Panel Layout designing with the name **Panel**; refer to Figure-3. The procedures of using various tools of the panel are discussed next.

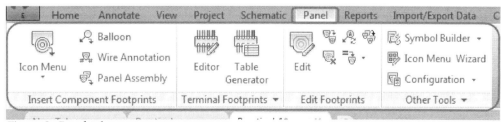

Figure-3. Panel tab

ICON MENU

The **Icon Menu** tool works in a similar way as it does for Schematic drawings. The only difference is between its representation of components in the panel layout. The procedure to use this tool is given next.

- Click on the **Icon Menu** tool from the **Insert Component Footprints** drop-down in the **Insert Components Footprints** panel in the **Panel** tab of the **Ribbon**; refer to Figure-4. The Insert Footprint dialog box will be displayed; refer to Figure-5.

Figure-4. Icon Menu tool

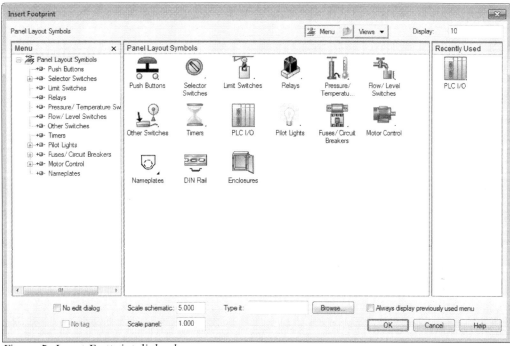

Figure-5. Insert Footprint dialog box

- Click on the desired category (Push Buttons in our case). The symbols related to the selected category will be displayed; refer to Figure-6.

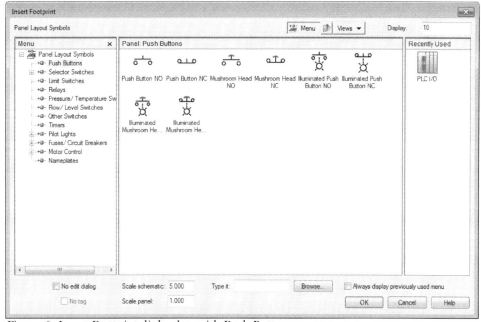

Figure-6. Insert Footprint dialog box with Push Buttons

- Click on the desired symbol from the dialog box (**Push Button NC** in our case). The **Footprint** dialog box will be displayed; refer to Figure-7.

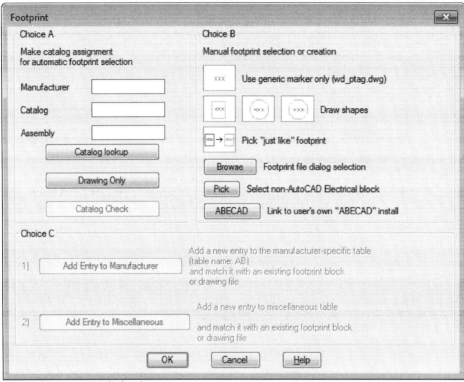

Figure-7. Footprint dialog box

- Click on the **Catalog lookup** button to check the catalog for current symbol. The **Catalog Browser** dialog box will be displayed; refer to Figure-8.

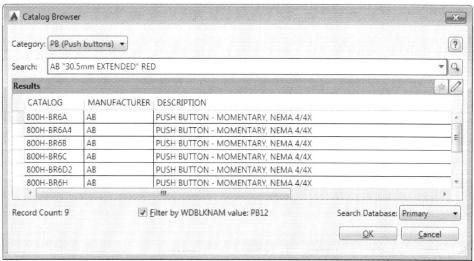

Figure-8. Catalog Browser dialog box

- Select the desired component from the catalog and click on the **OK** button from the dialog box.
- If you have created a block for the component in your local drive then click on the **Browse** button and select the desired component block from the dialog box; refer to Figure-9.

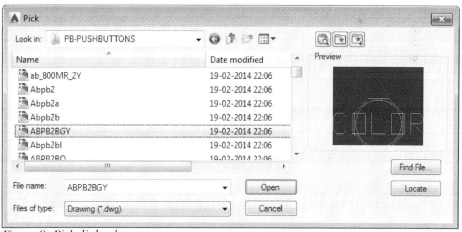

Figure-9. Pick dialog box

- After selecting the desired component, click on the Open button from the dialog box. The symbol will get attached to the cursor.
- Click at the desired place in the drawing to place the

symbol. You are asked to specify the rotation angle for the symbol.
- Specify the desired angle or press ENTER to use the default angle (i.e. 0 degree). The **Panel Layout-Component Insert/Edit** dialog box will be displayed; refer to Figure-10.

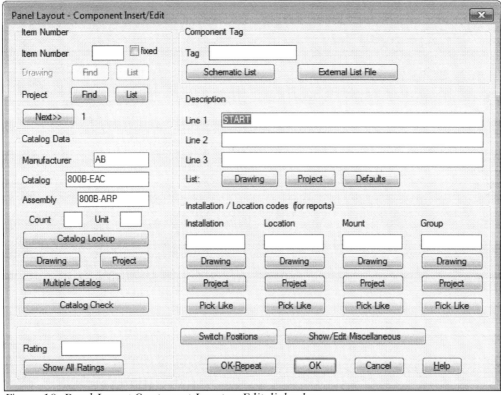

Figure-10. Panel Layout Component Insert or Edit dialog box

- This dialog box is similar to the **Insert/Edit Component** dialog box as discussed earlier. Specify the tags for the component.
- Click on the **Schematic List** button below **Tag** edit box in the **Component Tag** area to display the list of tags used in the drawing/project. The **Schematic Tag List** dialog box will be displayed; refer to Figure-11.
- Select the tag related to the symbol being inserted and click on the **OK** button from the dialog box.
- Click on the **OK** button from the dialog box. The symbol will be placed with the corresponding tag attached. If you want to insert more instances of the component then

click on the **OK-Repeat** button from the dialog box.

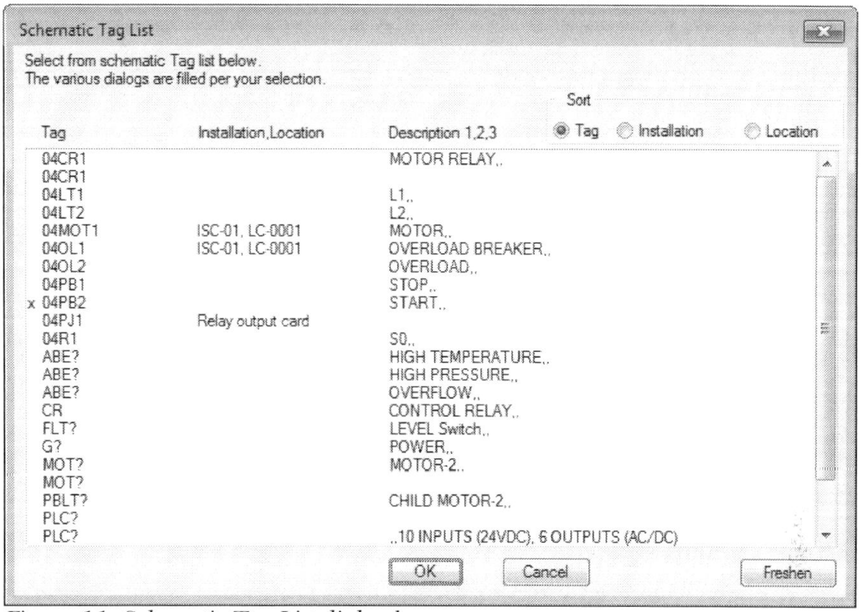

Figure-11. Schematic Tag List dialog box

Before we start inserting components randomly, it is important to understand the basic frame of panels. An electric panel generally has an encloser to close pack all the components of panel, a DIN rail to support the MCBs(Mounted Circuit Breakers), Nameplates to identify components, and electrical components. We will create a panel combining all of them later in this chapter (under Practical).

SCHEMATIC LIST

The **Schematic List** tool is used to import the list of schematic components from the current project drawings so that the panel representation of those components can be created. The procedure to use this tool is given next.

- Click on the **Schematic List** tool from the **Insert Component Footprints** drop-down. The **Schematic Components List** dialog box will be displayed; refer to Figure-12.

Figure-12. Schematic Components List dialog box

- Select the **Active drawing** if you want to extract the component list of only current drawing (as in our case) or select the **Project** and include the relevant drawings.
- Click on the **OK** button from the dialog box. The **Schematic Components (active drawing)** dialog box will be displayed (in case of selecting **Active drawing** radio button); refer to Figure-13.

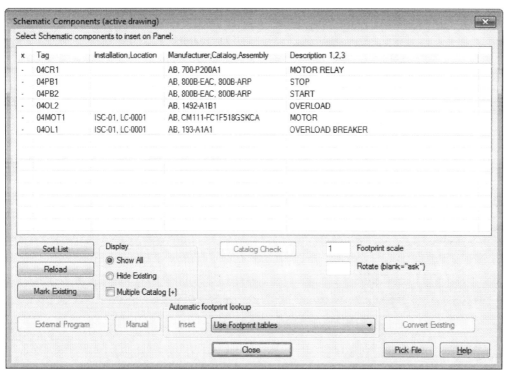
Figure-13. Schematic Components dialog box

- Select any component from the list for which you want to insert symbol in the panel. Click on the **Insert** button at the bottom of the dialog box. If the symbol for the component then it will get attached to the cursor and you will asked to specify the insertion point for the symbol.
- Click in the drawing area to place the component. You will be asked to specify the rotation angle value for the component.
- Specify the desired value in the Command box or press ENTER to set the default angle. The **Panel Layout-Component Insert/Edit** dialog box will be displayed as discussed earlier. Note that the tags are defined automatically this time. This is because the tags are automatically extracted as per the schematic list.
- If you want to perform modifications then do that and click on the **OK** button from the dialog box. The symbol will be created and the **Schematic Components** dialog box will appear again.
- Follow the same procedure and insert all the components required in the panel.
- After inserting all the components, click on the **Close** button from the dialog box to exit.

MANUAL

The **Manual** tool is used to insert the panel components as per our requirement directly using the blocks. In this case, we need to be very careful regarding the tags otherwise it can create great problems for assembly site. The procedure to use this tool is given next.

- Click on the **Manual** tool from the **Insert Component Footprints** drop-down. The **Insert Component Footprints -- Manual** dialog box will be displayed; refer to Figure-14.

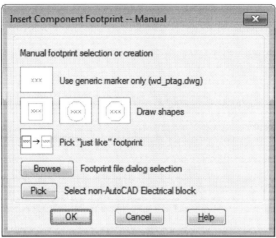

Figure-14. Insert Component Footprint Manual

- This dialog box is similar to the Choice B area of the Footprint dialog box discussed earlier. Click on the Browse button to insert the block of component.
- If you have the component already in the drawing, then click on the **Pick** button and select the component. (Note that you can select only non-AutoCAD Electrical blocks as symbol for using this feature).
- On selecting the block, you are asked to specify the insertion point for the symbol.
- Click in the drawing area and specify the rotation of the symbol. The symbol will be placed and **Panel Layout - Component Insert/Edit** dialog box will be displayed as discussed earlier.
- Specify the desired tags and click on the **OK** button from the dialog box.

MANUFACTURER MENU

The **Manufacturer Menu** tool is used to insert the components from the manufacturer's menu. AutoCAD Electrical contains a library of manufacturer's components (**Allen-Bradley** and **ABECAD**). The procedure to use this tool is given next.

- Click on the **Manufacturer Menu** tool from the **Insert Component Footprints** drop-down. The **Vendor Menu Selection** dialog box will be displayed (for the first time); refer to Figure-15.

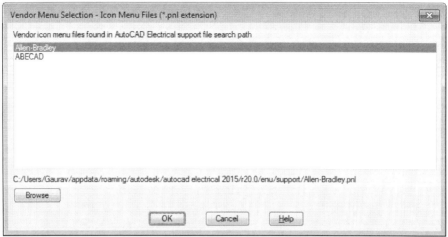

Figure-15. Vendor Menu Selection dialog box

- Select the desired vendor from the list and click on the **OK** button from the dialog box. The **Vendor Panel Footprint** dialog box will be displayed; refer to Figure-16.

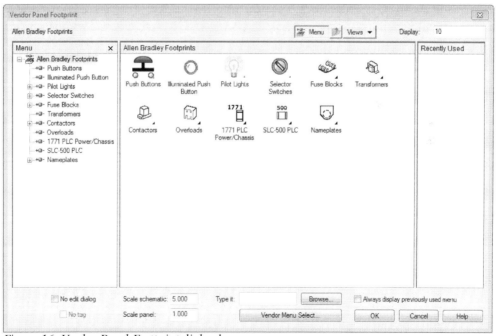

Figure-16. Vendor Panel Footprint dialog box

- The dialog box works in similar way to **Insert Component** dialog box discussed earlier.

Similarly, you can use the **User Defined List** and **Equipment List** tools to insert the **Panel** components.

BALLOON

The **Balloon** tool is used to assign balloons to the components for their identification in the table. The procedure to use this tool is given next.

- Click on the **Balloon** tool from the **Insert Component Footprints** panel in the **Ribbon**. You are asked to select the component for which you want to assign the balloon or specify settings for the balloon.
- Enter **S** at the command prompt to specify parameters related to balloon. The **Panel balloon setup** dialog box will be displayed; refer to Figure-17.

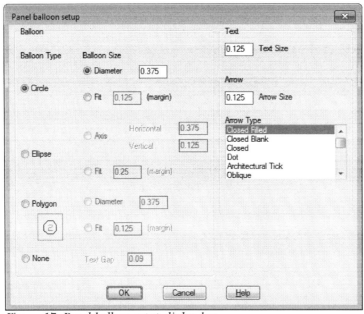

Figure-17. Panel balloon setup dialog box

- Select the desired balloon type and specify the parameters in the dialog box.
- Click on the **OK** button from the dialog box to apply the specified parameters.
- Click on the component. You are asked to specify the leader start point or balloon insertion point.

- Click at the desired point near the component. A dashed line of leader will be displayed having end point attached to the cursor.
- Click to specify the end point of the leader or press **ENTER** to place the balloon at the earlier selected point; refer to Figure-18.

Figure-18. Component with assigned balloon

- If you have specified the end point of the leader then press **ENTER** to place the balloon.

WIRE ANNOTATION

The **Wire Annotation** tool is used to attach wire annotations to the selected panel components in the drawing. The procedure to use this tool is given next.

- Click on the **Wire Annotation** tool from the **Insert Component Footprints** panel. The **Schematic Wire Number --> Panel Wiring Diagram** dialog box will be displayed; refer to Figure-19.
- Select the desired radio button from the **Panel connection annotation for** area. If you select the **Active drawing(all)** then the annotations will be assigned to components in the current drawing based on all the active drawings.

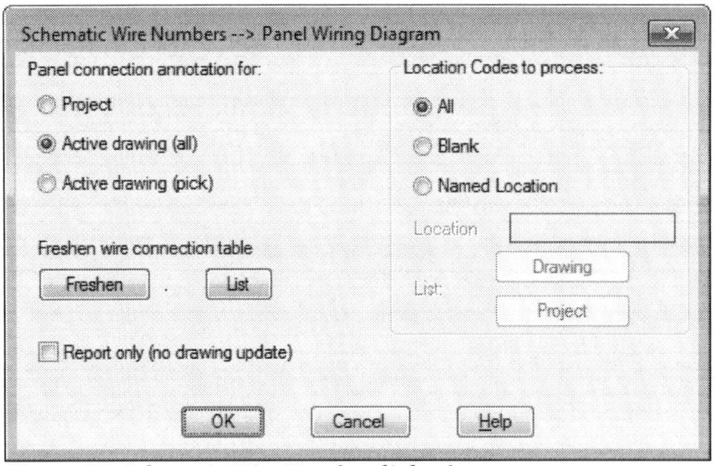

Figure-19. Schematic Wire Number dialog box

- Select the desired radio button from the **Location Codes to process** area. The location codes will be assigned to the components as per the selection of the radio button.
- After specifying the desired settings, click on the **OK** button from the dialog box. The **Schematic --> Layout Wire Connection Annotation** dialog box will be displayed; refer to Figure-20.

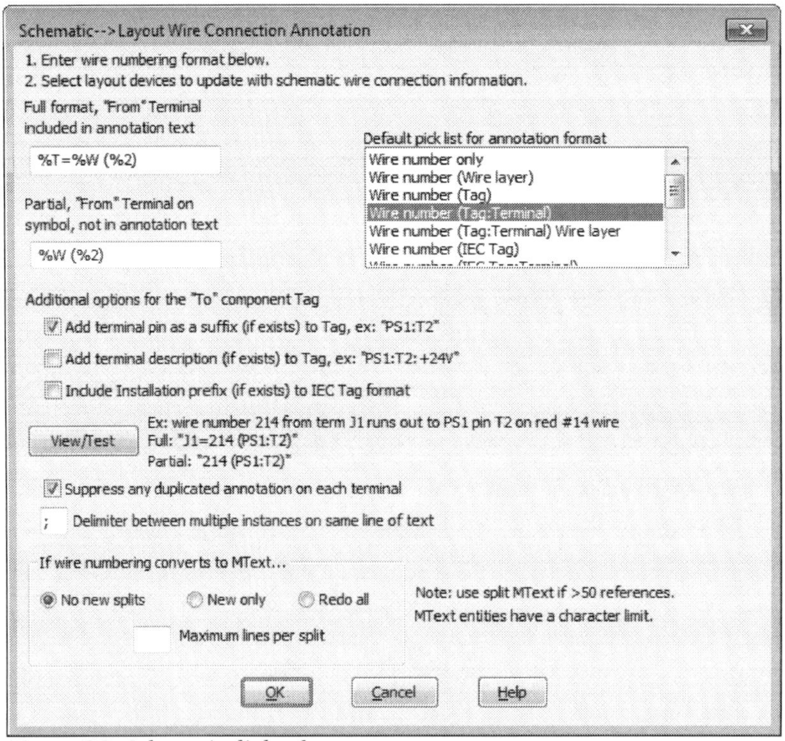

Figure-20. Schematic dialog box

- Specify the annotation format in this dialog box and click on the OK button from the dialog box. The annotations will be assigned to the components in the current drawing automatically; refer to Figure-21.

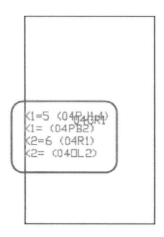

Figure-21. Annotations automatically assigned

PANEL ASSEMBLY

The **Panel Assembly** tool is used to insert an already created assembly of panel in the current drawing. This tool is very useful if you are going to use the same type of panel assemblies in various drawings. The procedure to use this tool is given next.

- Click on the **Panel Assembly** tool from the **Insert Component Footprints** panel in the **Ribbon**. The **Insert Panel Assembly of Blocks** dialog box will be displayed; refer to Figure-22.

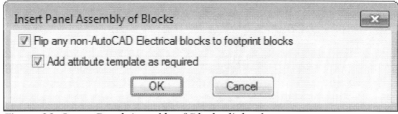

Figure-22. Insert Panel Assembly of Blocks dialog box

- Make sure all the check boxes are selected and then click on the **OK** button from the dialog box. The **WBlocked Assembly to insert** dialog box will be displayed; refer to Figure-23.

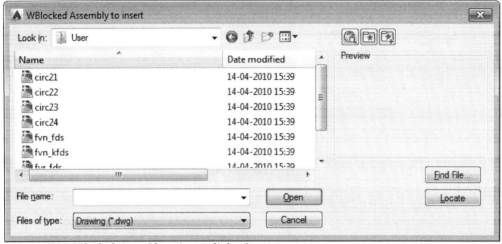

Figure-23. WBlocked Assembly to insert dialog box

- Select the drawing file of earlier created panel block and click on the **Open** button from the dialog box. You are asked to specify the insertion point for the panel block.
- Click in the drawing area to specify the insertion point for the panel block. You are asked to specify the rotation angle value for the block.
- Specify the desired angle value and press ENTER. The block will be placed.

Till this point, we have discussed about inserting the component footprints in the panel layout. Now, we will learn to terminal footprints. The tools to manage the terminal footprints are available in the **Terminal Footprints** panel in the **Ribbon**. The tools in this panel are discussed next.

EDITOR

The **Editor** tool is used to edit the details of terminals present in the current drawing. Note that the terminal information is collected from all the drawings in the current project. So, it is important to activate the project whose drawing is being used for editing terminal footprints. The procedure to use the **Editor** tool is given next.

- Click on the **Editor** tool from the **Terminal Footprints** panel. The **Terminal Strip Selection** dialog box will be displayed; refer to Figure-24.

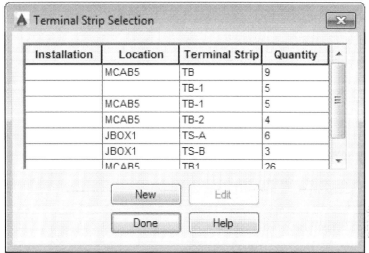

Figure-24. Terminal Strip Selection dialog box

- Select the entry that you want to edit in this table and click on the **Edit** button. If the number of terminals in the current drawing exceed the number of wires then the **Defined Terminal Wiring Constraints Exceeded** dialog box will be displayed; refer to Figure-25.

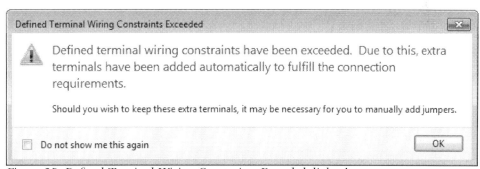

Figure-25. Defined Terminal Wiring Constraints Exceeded dialog box

- Click on the **OK** button from the dialog box. The **Terminal Strip Editor** dialog box will be displayed; refer to Figure-26.

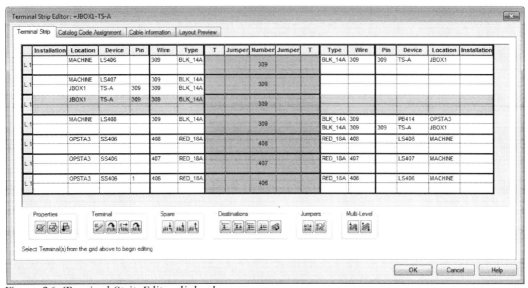

Figure-26. Terminal Strip Editor dialog box

- The selected terminal will be highlighted in yellow color. There are six areas in this dialog box named; **Properties**, **Terminal**, **Spare**, **Destinations**, **Jumpers**, and **Multi-Level**. The tools in these areas are used to edit the respective parameters of the terminals. Hold the cursor on each of the button in these areas to check its function.
- After specifying the desired parameters click on the **OK** button from the dialog box. If you have changed the graphical parameters then the **Graphical Terminal Strip Insert/Update Required** dialog box will be displayed; refer to Figure-27.

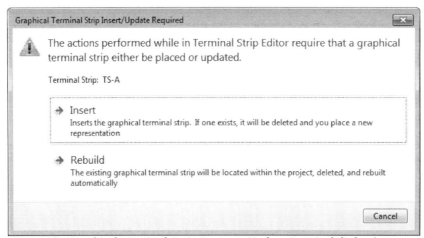

Figure-27. Graphical Terminal Strip Insert or Update Required dialog box

- Click on the **Insert** button if you want to insert new graphical representation of the terminal strip. The strip will get attached to the cursor; refer to Figure-28.

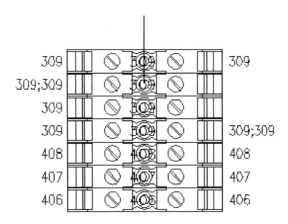

Figure-28. New terminal strip attached to cursor

- In place of selecting the **Insert** button, you can select the **Update** button to update the already existing terminal strip.
- After updating or inserting the strip, the **Terminal Strip Selection** dialog box will be displayed again.
- If you want to add a new terminal strip, then click on the **New** button from the dialog box. The **Terminal Strip Definition** dialog box will be displayed; refer to Figure-29.

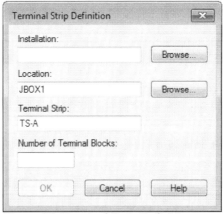

Figure-29. Terminal Strip Definition dialog box

- Click in the **Number of Terminal Blocks** edit box and specify the number of blocks of terminals. The **OK** button will become active.
- Click on the **OK** button from the dialog box. The **Terminal Strip Editor** dialog box will be displayed as discussed earlier. Specify each detail of the terminal as required and then click on the **OK** button from the dialog box. The dialog box will be displayed asking you to insert the graphical representation of the terminal.
- Click on the **Insert** button from the dialog box and place the terminal at the desired position in the drawing.
- Click on the **Done** button from the **Terminal Strip Selection** dialog box to exit.

TABLE GENERATOR

The **Table Generator** tool is used to insert a table consisting of details related to the terminals in the drawing. The procedure to use this tool is given next.

- Click on the **Table Generator** tool from the **Terminal Footprints** panel in the **Ribbon**. The **Terminal Strip Table Generator** dialog box will be displayed; refer to Figure-30.

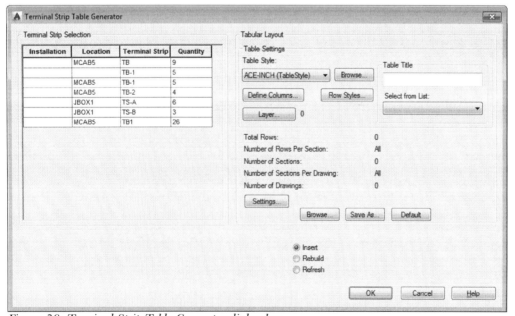

Figure-30. Terminal Strip Table Generator dialog box

- Select the desired terminal from the **Terminal Strip Selection** table in the dialog box. For multiple selection, press and hold the CTRL key and then select; refer to Figure-31.

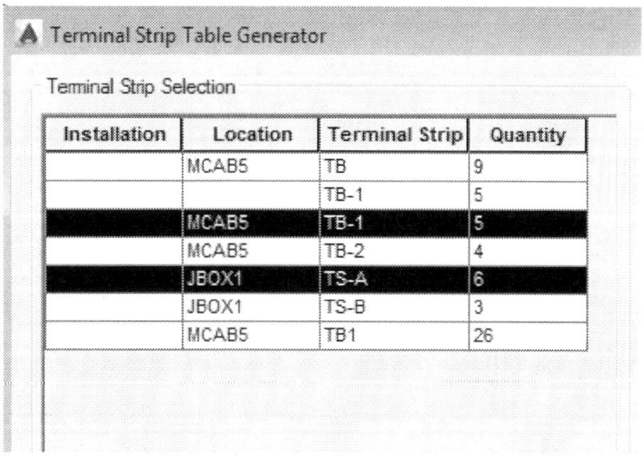

Installation	Location	Terminal Strip	Quantity
	MCAB5	TB	9
		TB-1	5
	MCAB5	TB-1	5
	MCAB5	TB-2	4
	JBOX1	TS-A	6
	JBOX1	TS-B	3
	MCAB5	TB1	26

Figure-31. Terminal Selection

- Click on the **Table Style** drop-down in the **Table Settings** area of the dialog box and select the desired table style.
- Click in the **Table Title** edit box and specify the suitable title for the terminal table.
- Click on the **Settings** button in the dialog box. The **Terminal Strip Table Settings** dialog box will be displayed; refer to Figure-32.
- Click on the **Browse** button for the **First Drawing Name** edit box in the **Drawing Information for Table Output** area of the dialog box. The **First Drawing Name** dialog box will be displayed; refer to Figure-33.
- Select the desired drawing(The name of drawing in which table will be inserted is decided on the basis of the selected drawing's name). Click on the **Save** button from the dialog box.
- After specifying the rest of the parameters as required, click on the **OK** button from the dialog box.

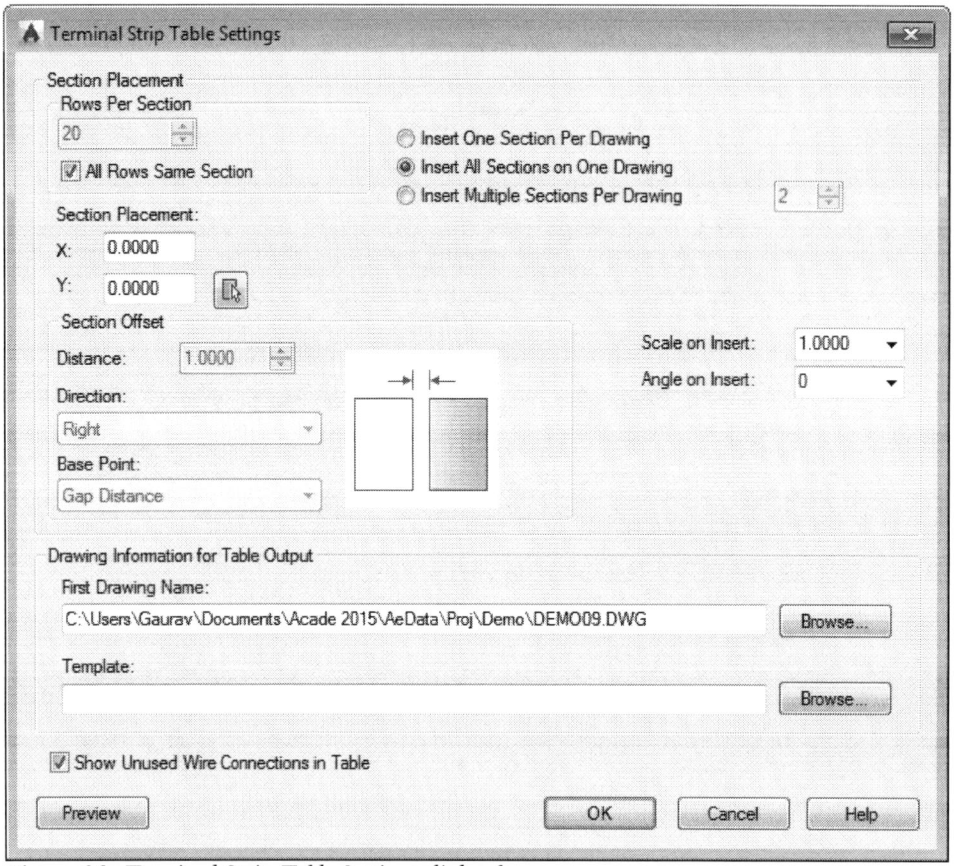

Figure-32. Terminal Strip Table Settings dialog box

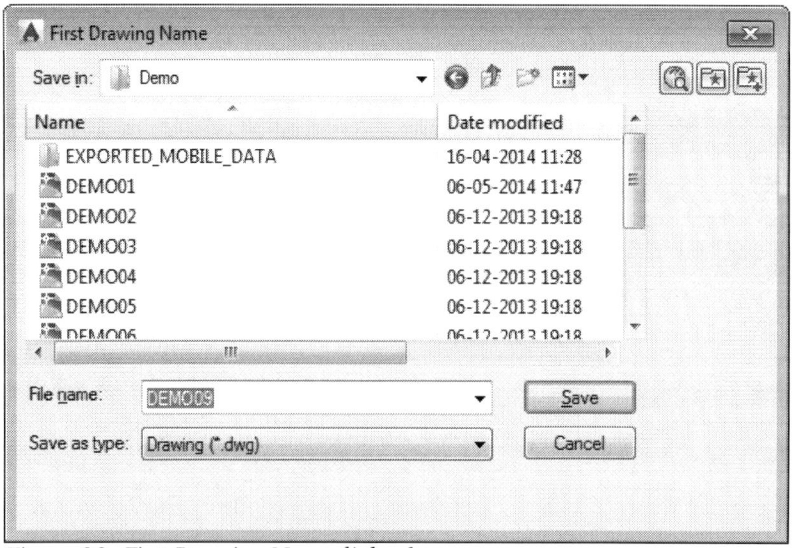

Figure-33. First Drawing Name dialog box

- Click on the **OK** button from the **Terminal Strip Table Generator** dialog box. The **Table(s) Inserted** dialog box will be displayed; refer to Figure-34.

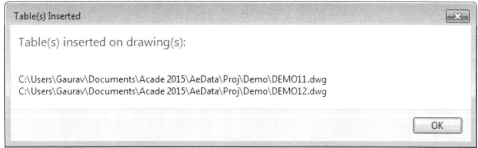
Figure-34. Tables Inserted dialog box

- Click on the **OK** button from the dialog box. Drawing will be added in the current project having tables created recently; refer to Figure-35.

Figure-35. Table created

INSERT TERMINALS

There are two tools to insert terminals in the panel layout; **Insert Terminals (Schematic List)** and **Insert Terminals (Manual)** which work similar to **Schematic List** and **Manual** tool respectively. These tools are available in the **Insert Component Footprints** drop-down in the **Insert Component Footprints** panel. Please follow the procedures given for these tools to insert terminals.

EDITING FOOTPRINTS

The tools to edit footprints are available in the **Edit Footprints** panel; refer to Figure-36. These tools are discussed next.

Figure-36. Edit Footprints panel

Edit

The **Edit** tool is used to modify the footprints of components in the panel layout. The procedure to use this tool is given next.

- Click on the **Edit** tool from the **Edit Footprints** panel. You are asked to select the component for editing footprint.
- Select the desired component. The **Panel Layout - Component Insert/Edit** dialog box will be displayed; refer to Figure-37.

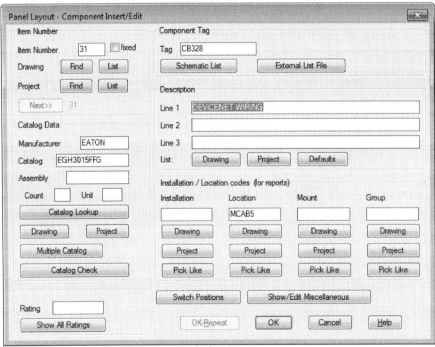

Figure-37. Panel Layout Component dialog box

- Specify the desired tags and parameters in the dialog box and click on the **OK** button from the dialog box to apply the specified settings.

Copy Footprint

The **Copy Footprint** tool is used to copy the footprint of a component so that you can assign it to other components in the panel layout. The procedure is given next.

- Click on the **Copy Footprint** tool from the **Edit Footprints** panel. You are asked to select the components just alike the component to which you want to assign footprints.
- Select the desired component. Its footprint will get attached to the cursor; refer to Figure-38.

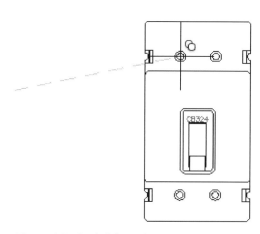

Figure-38. Copied footprint

- Click at the desired location to place the footprint. The **Panel Layout - Component Insert/Edit** dialog box will be displayed. The options in this dialog box have already been discussed.
- Follow the procedures discussed earlier.

Delete Footprint

The **Delete Footprint** tool is used to delete the footprints of panel components. The procedure is given next.

- Click on the **Delete Footprint** tool from the **Edit Footprints** panel. You are asked to select the components whose footprints are to be deleted.
- One by one click on the components that you want to be deleted and press ENTER. The footprints will be deleted and the **Search for / Surf to Children** dialog box will be displayed; refer to Figure-39.

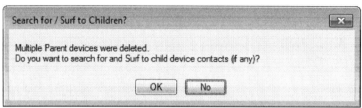

Figure-39. Search for or Surf to Children dialog box

- Click on the **OK** button if you want to delete the children components also or click on the No button to exit the tool.
- If you clicked on the **OK** button then the **Surf** dialog box will be displayed; refer to Figure-40.

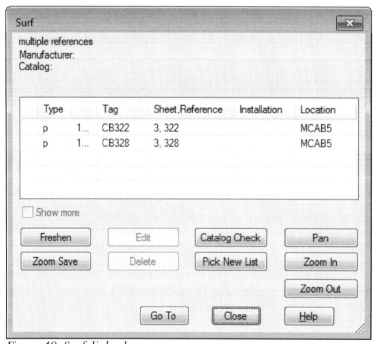

Figure-40. Surf dialog box

- Select the item from the list and click on the **Go To** button. The component will be displayed in the corresponding drawing and the **Delete** button will become active; refer to Figure-41.

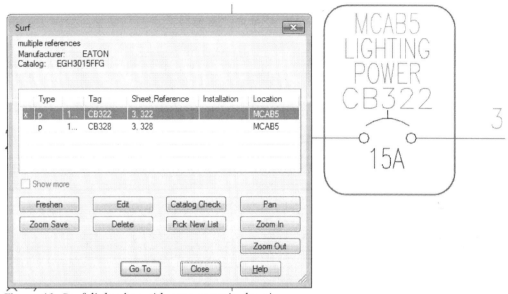

Figure-41. Surf dialog box with component in drawing

- Click on the **Delete** button and delete the children components if you want to.

Resequence Item Numbers

The **Resequence Item Numbers** tool is used to re-sequence all the items in the current drawing/project. The procedure is given next.

- Click on the **Resequence Item Numbers** tool from the **Edit Footprints** panel. The **Resequence Item Numbers** dialog box will be displayed; refer to Figure-42.
- Specify the desired starting number for the items in the Start edit box.
- Select the desired radio button and click on the **OK** button from the dialog box. (In our case, **Active drawing(all)** radio button is selected).
- On doing so, the item numbers will be updated automatically.

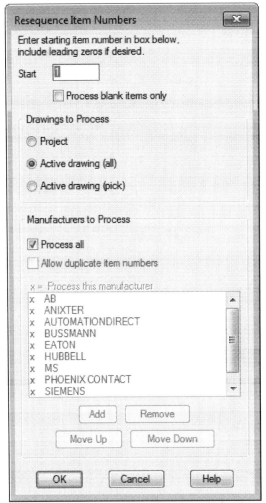

Figure-42. Resequence Item Numbers dialog box

Copy Codes drop-down

There are four tools in this drop-down; **Copy Installation Code**, **Copy Location Code**, **Copy Mount Code**, and **Copy Group Code**. These tools work in the same way. The procedure for **Copy Location Code** tool is given next. You can assume the same for other tools in the drop-down.

- Click on the **Copy Location Code** tool from the **Copy Codes** drop-down. The **Copy Installation\Location\Mount\Group to Components** dialog box will be displayed; refer to Figure-43.

Figure-43. Copy Installation Location Mount Group to Components dialog box

- Click on the **Pick Master** button. You are asked to select the component whose location code is to be copied.
- Select the desired component. The dialog box will be displayed again. Click on the **OK** button from the dialog box. You are asked to select the target components.
- Select the target components to assign the copied location code and press ENTER. The location code will be assigned.

Copy Assembly

The **Copy Assembly** tool is used to copy one or more components of a panel assembly. The procedure to use this tool is given next.

- Click on the **Copy Assembly** tool from the drawing. You are asked to select components or assembly.
- Select the desired components and press ENTER. You are asked to specify the base point for copying.
- Click to specify the base point. The component will get attached to the cursor.
- Click in the drawing area to place the copied components/assembly.

Practical 1

Create the panel layout for schematic diagram created in Practical 1 of Chapter 8; refer to Figure-44.

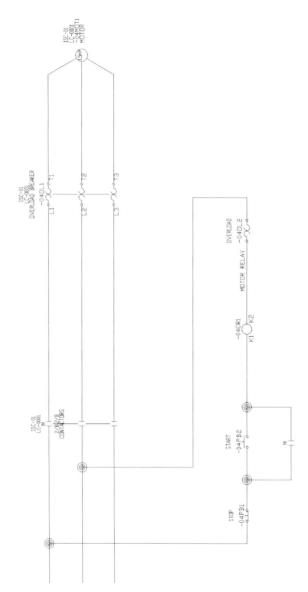

Figure-44. Practical 1 schematic

Starting a drawing

- Select a project having schematic drawing for which you want to design the panel. Note that you need to add the drawing of Practical 1 of previous chapter in any project.

- Right click on the project name in the **Project Manager**; refer to Figure-45.

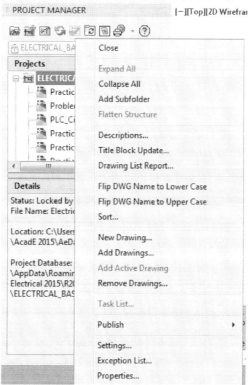

Figure-45. Shortcut menu of Project

- Click on the **New Drawing** button. The **Create New Drawing** dialog box will be displayed; refer to Figure-46.
- Click in the **Name** edit box and specify the name as **Panel Layout**.
- Click on the **Drawing** button and select the code for **Installation** as well as **Location** codes.

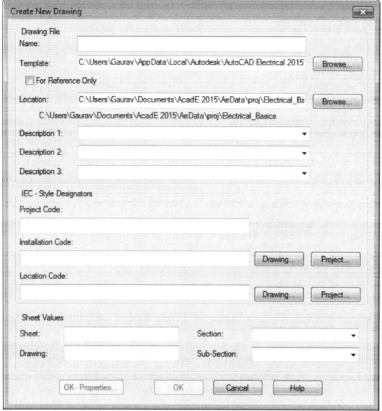

Figure-46. Create New Drawing dialog box

- Click on the **OK** button from the dialog box. The **Apply Project Defaults to Drawing Settings** dialog box will be displayed; refer to Figure-47.

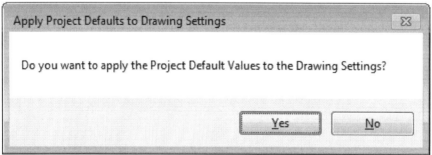

Figure-47. Apply Project Defaults to Drawing Settings dialog box

- Click on **Yes** to apply the settings. The drawing will be created; refer to Figure-48.

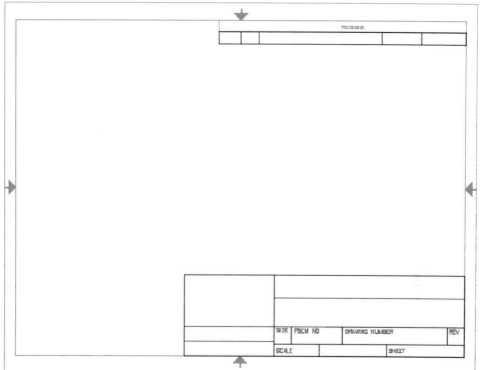

Figure-48. Drawing created

Adding Enclosure of panel

- Click on the **Icon Menu** button from the **Insert Component Footprints** drop-down in the **Insert Component Footprints** panel. For the first time, the **Alert** dialog box will be displayed; refer to Figure-49.

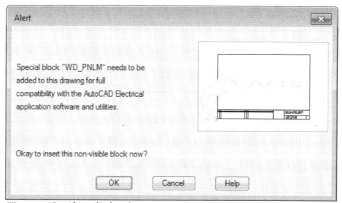

Figure-49. Alert dialog box

- Click on the **OK** button from the dialog box. The **Insert Footprint** dialog box will be displayed; refer to Figure-50.

Figure-50. Insert Footprint dialog box

- Click on the **Enclosures** category in the dialog box. The **Footprint** dialog box will be displayed; refer to Figure-51.
- Click on the **Catalog lookup** button to find out the manufacturer data or specify the desired manufacturer data in edit boxes of the **Choice A** area.
- Click on the **Browse** button and select the drawing block for the enclosure.
- Click on the **Open** button from the dialog box, you are asked to specify the insertion point for the footprint.
- Click in the drawing area to specify the insertion point. You are asked to specify the rotation angle for the current footprint. Enter **0** at the command prompt. The enclosure will be placed; refer to Figure-52 and the **Panel Layout** dialog box will be displayed; refer to Figure-53.

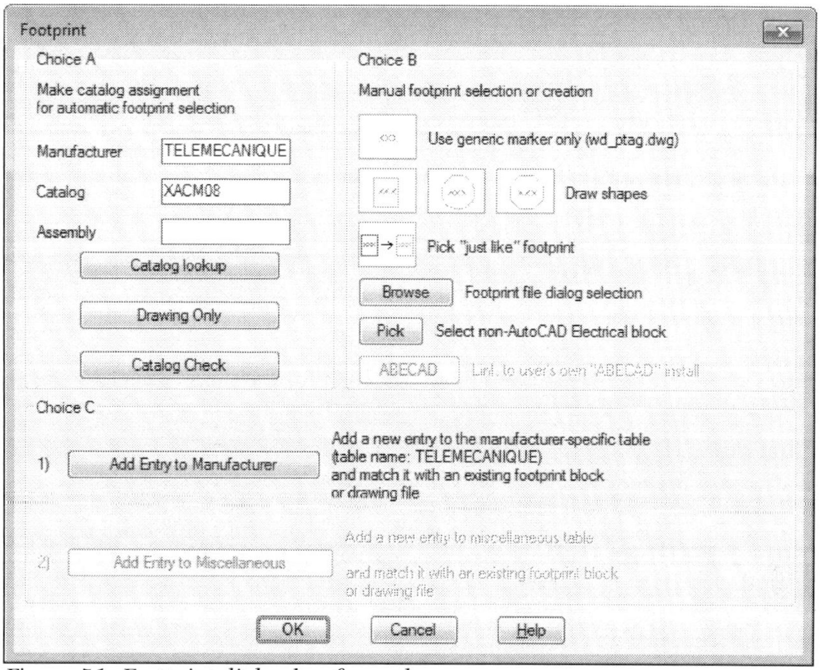

Figure-51. Footprint dialog box for enclosure

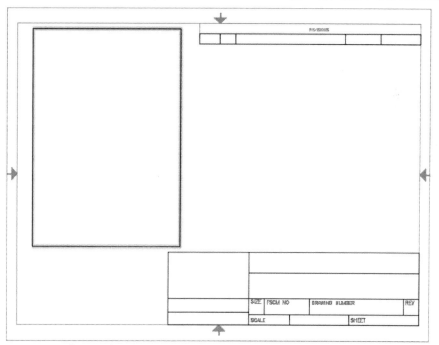

Figure-52. Enclosure placed

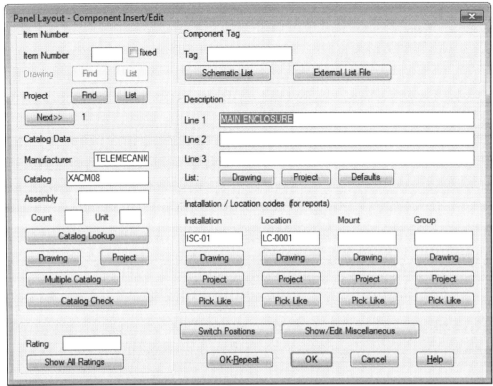

Figure-53. Panel Layout dialog box for enclosure

- Make sure that you have the values specified as in the above figure and then click on the **OK** button from the dialog box.

Inserting other components of the panel

- Click on the **Schematic List** tool from the **Insert Component Footprints** drop-down in the **Insert Component Footprints** panel. The **Schematic Components List** dialog box will be displayed; refer to Figure-54.

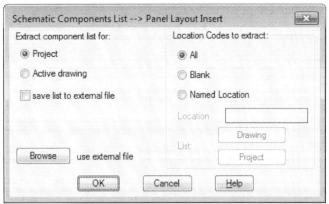

Figure-54. Schematic Components List dialog box

- Make sure that the **Project** radio button is selected and then click on the **OK** button from the dialog box. The **Select Drawings to Process** dialog box will be displayed; refer to Figure-55.

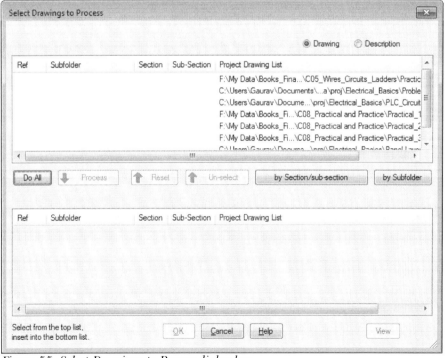

Figure-55. Select Drawings to Process dialog box

- Select the **Practical 1** drawing from the list and click on the **Process** button. The drawing will be included in the schematic list; refer to Figure-56.

Figure-56. Drawing included for processing

- Click on the **OK** button. The **Schematic Components (active project)** dialog box will be displayed; refer to Figure-57.

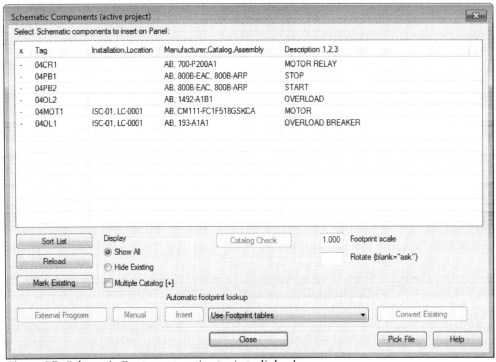

Figure-57. Schematic Components active projects dialog box

- Select the component with description STOP and click on the Insert button. The icon will get attached to the cursor; refer to Figure-58.

Figure-58. Stop button attached to cursor

- Click in the enclosure to place the button at the desired place. You are asked to specify the rotation angle.
- Enter **0** at the command prompt. The component will be placed and the **Panel Layout - Insert/Edit Component** dialog box will be displayed.
- Specify the required parameters and click on the **OK** button. The button will be placed and the **Schematic Components (active project)** dialog box will be displayed again.
- Similarly insert the other components of the user panel and assign balloons as discussed in the chapter. After adding all the components and the balloons the panel layout will be displayed as shown in Figure-59.

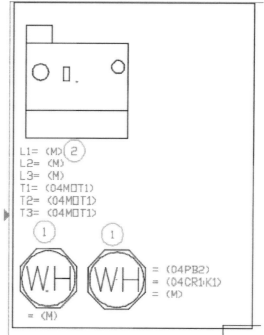

Figure-59. Panel drawing for Practical 1

PRACTICE 1

Create a panel layout for users for the schematic created in Practical 2 of chapter 8; refer to Figure-60. The panel layout is given in .

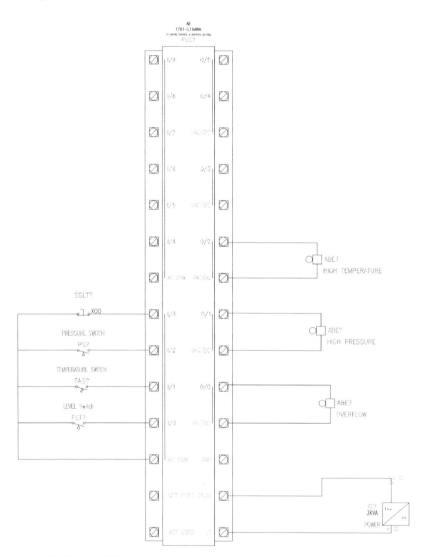

Figure-60. Practical 3

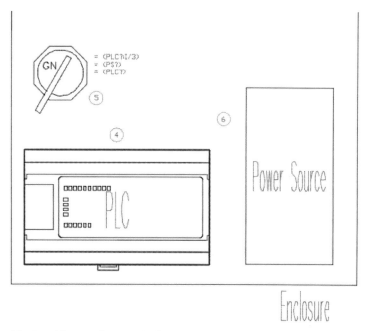

Figure-61. Panel layout for practical 3

Page left blank intentionally for student notes

Reports

Chapter 10

The major topics covered in this chapter are:

- *Introduction*
- *Reports*
- *Missing Data Catalog*
- *Electrical Audit*
- *Drawing Audit*
- *Dynamic Table editing*

Introduction

After the huge discussion of technical aspects in previous chapters, this is the time to relax and see the results of your work. Actually reports are the tabulated representation of components being used in the schematic diagrams and panel layouts. The tools to create reports are available in the **Reports** tab of the **Ribbon**; refer to Figure-1. Before we start using the tools one by one, it is important to note that the reports created here are totally dependent on the schematic diagrams and panel layouts created earlier. The tables generally have Names of components, Manufacturer details of components and quantity of components. We can include other details also like location code, installation code and so on in the table. The tools to create tables are discussed next.

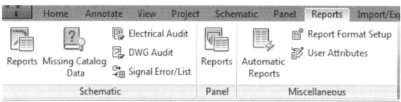

Figure-1. Reports panel

REPORTS (SCHEMATIC)

The **Reports** tool in the **Schematic** panel is used to create reports of schematic components of the project. The procedure to use this tool is given next.

- Click on the **Reports** tool from the **Schematic** panel in the **Reports** tab. The **Schematic Reports** dialog box will be displayed; refer to Figure-2.

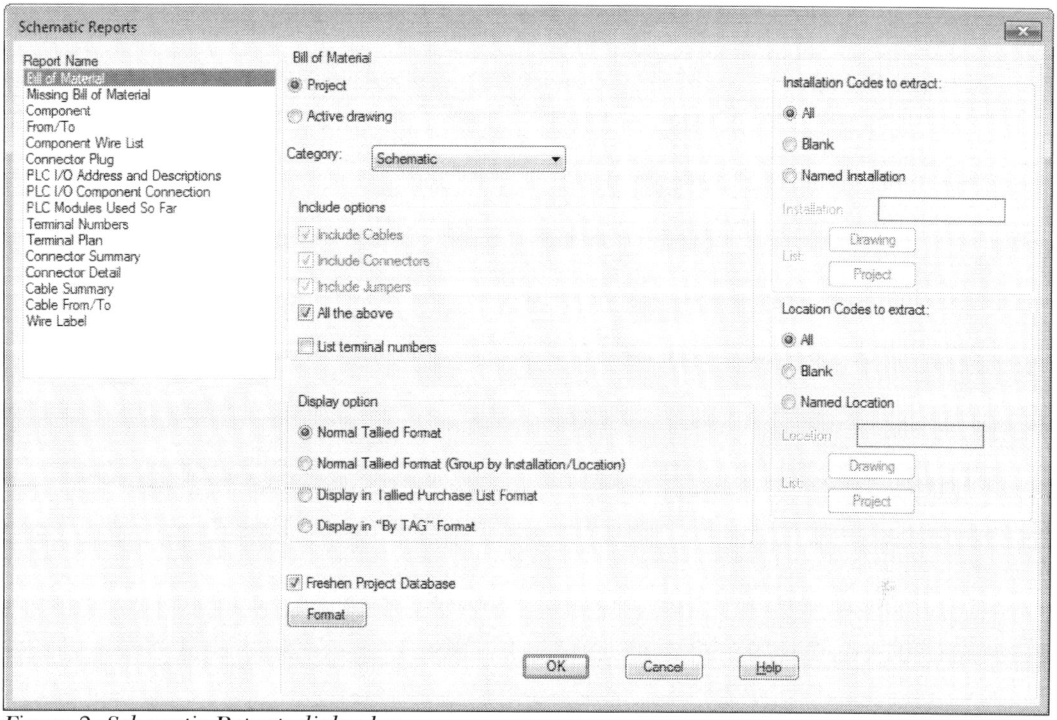

Figure-2. Schematic Reports dialog box

- We have a big list of report types in the **Report Name** selection box at the left; refer to Figure-3. Select the desired report type from the selection box, the options in the dialog box will be modified accordingly.

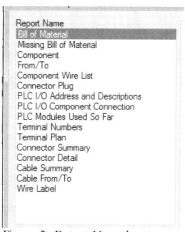

Figure-3. Report Name box

- The list of report types available in the dialog box is given next:

Bill of Materials reports

The Bill of Material reports report only components with assigned BOM information. These reports provide the following BOM-related features:
- Extract BOM reports on demand, active drawing, or project-wide
- Extract BOM reports on a per-location basis
- Change BOM report format
- Output BOM reports to ASCII report file
- Export BOM data to a spreadsheet or database program
- Insert BOM as a table right on an AutoCAD drawing
- List parent or stand-alone components without catalog information

Procedure to create:
- Select the **Bill of Materials** option from the **Report Name** selection box. The options related to Bill of Materials will be displayed in the dialog box.
- Select the **Project** radio button if you want to include the drawings from the current project or select the **Active Drawing** radio button to create Bill of Materials based on the current active drawing.
- Click on the **Category** drop-down. The list of various options will be displayed; refer to Figure-4.

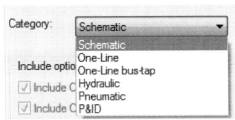

Figure-4. Category drop-down

- Select the desired option, the component related to the selected option will be included in the Bill of Materials report.
- Select the check boxes in the **Include options** area of the dialog box to include the corresponding components of the circuit in the report. You can select **All the above** check box to include all the items displayed above this check box.

- Select the **List terminal numbers** check box to include the terminal numbers in the report.
- The four radio buttons in the **Display option** area of the dialog box are used to specify sorting of the components in the table. The effect of these radio buttons is discussed next.

Normal Tallied Format

Identical component or component/assemblies are tallied and reported as single line items (example: Red push-button operator 800EP-F4 with 800E-A3L latch and two 800E-3X10 N.O. contact blocks).

Normal Tallied Format (Group by Installation/Location)

Identical component or component/assemblies with the same installation/location codes are tallied and reported as single line items.

Display in Tallied Purchase List Format

Each part becomes its own line item (example: no longer any sub-assembly items) and each is tallied across all component types. For example, all 800E-3X10 N.O. contact blocks for all components are reported as a single line item.

Display in By TAG Format

All instances of a given component-ID or terminal tag are processed together and then reported as a single entry.

- Select the desired radio button.
- Click on the Format button and select the desired format for the report.
- Select All radio buttons from **Installation Codes to extract** and **Location Codes to extract** areas to include installation codes and location codes in the report.
- Click on the **OK** button from the dialog box. The **Report Generator** dialog box will be displayed; refer to Figure-5.

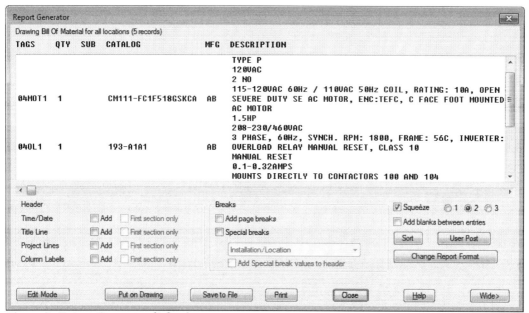

Figure-5. Report Generator dialog box

- Select the **Add** check box for desired data header from the **Header** area of the dialog box.
- Click on the **Edit Mode** button to edit the table data. The **Edit Report** dialog box will be displayed; refer to Figure-6.

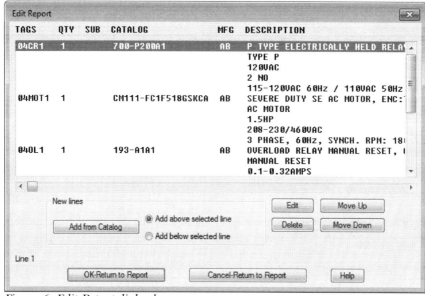

Figure-6. Edit Report dialog box

- Double-click on the desire row, the **Edit Line** dialog box will be displayed; refer to Figure-7.

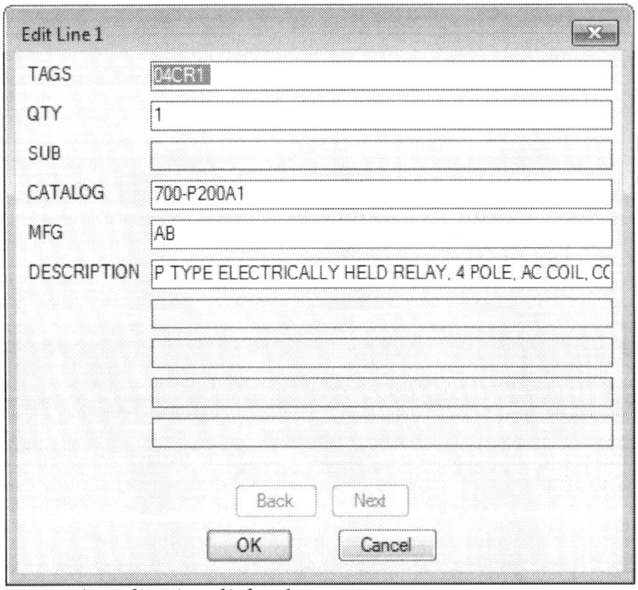

Figure-7. Edit Line dialog box

- Specify the desired values in the edit boxes of this dialog box and click on the **OK** button to exit the dialog box.
- You can add a new line by using the **Add from Catalog** button and use the catalog data.
- Click on the **OK-Return to Report** button to return the report creation.
- Now, we are ready to insert the report in the drawing. Click on the **Put on Drawing** button from the **Report Generator** dialog box. The **Table Generation Setup** dialog box will be displayed; refer to Figure-8.
- The options in this dialog box are used to modify the details of the table. Specify the desired parameters and click on the **OK** button from the dialog box. A box of table boundary will get attached to the cursor.

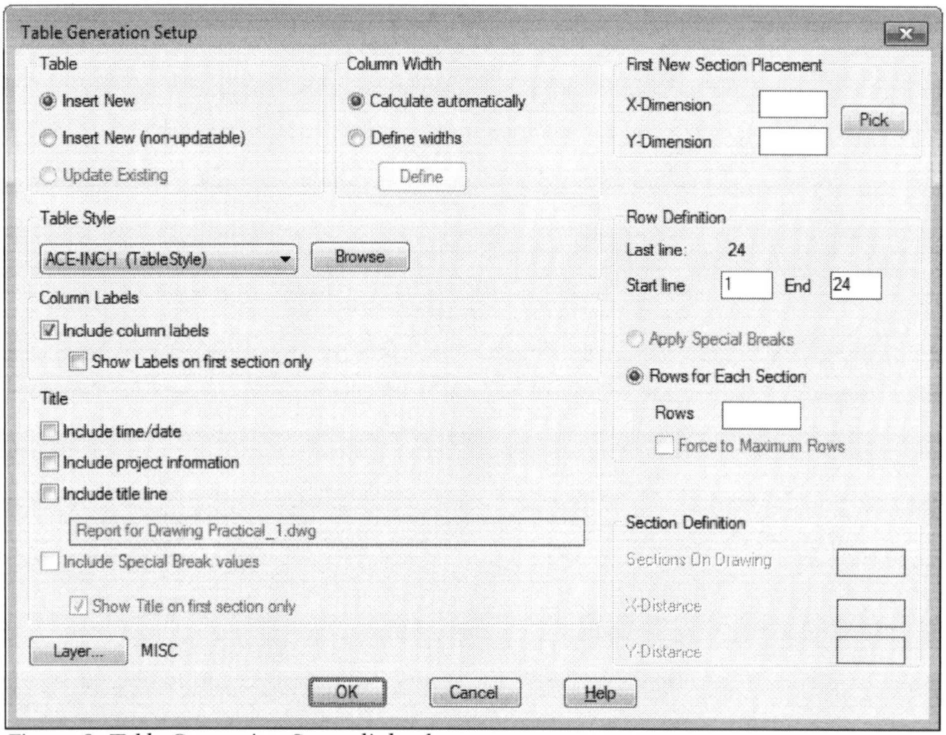

Figure-8. Table Generation Setup dialog box

- Click in at the desired place in the drawing to place the table. The table will be placed and the **Report Generator** dialog box will be displayed again.
- Click on the **Save to File** button if you want to save the report in a separate file. The **Save Report to File** dialog box will be displayed; refer to Figure-9.

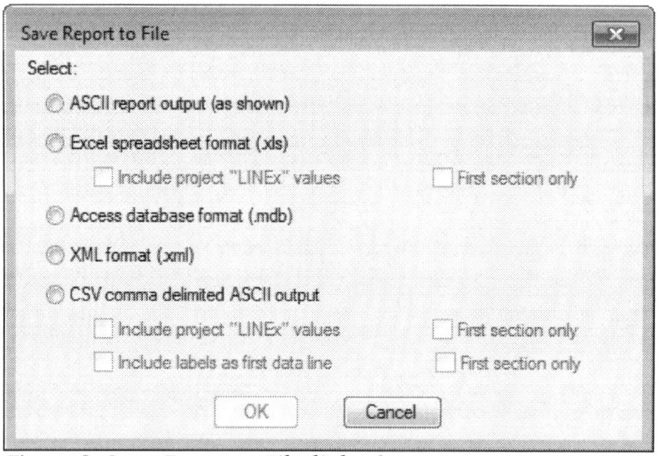

Figure-9. Save Report to File dialog box

- Select the desired radio button to specify the format for the file. (**Excel spreadsheet format(.xls)** in our case)
- Click on the **OK** button from the dialog box. The **Select file for report** dialog box will be displayed; refer to Figure-10.

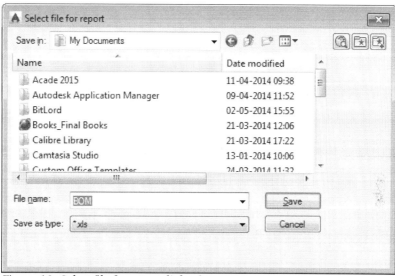

Figure-10. Select file for report dialog box

- Browse to the desired location in the hard-drive and save the file. On selecting the Save button, the **Optional Script File** dialog box will be displayed; refer to Figure-11.

Figure-11. Optional Script File dialog box

- Click on the **Run Script** button to generate the script file or click on the **Close-No Script** button to exit. If you generate the script file then it help the other programs to link the table data to its generation process.

- Click on the **Close** button from the **Report Generator** dialog box to exit. Figure-12 shows a table inserted in the drawing.

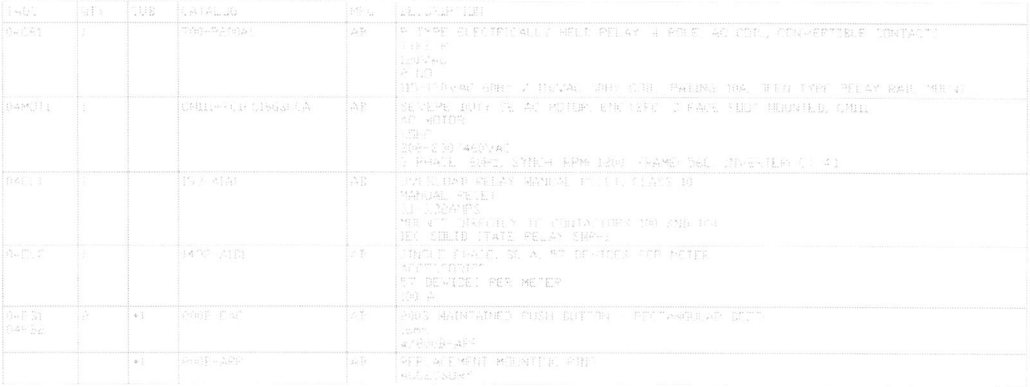

Figure-12. Report of BOM

Above procedure is similarly applicable to other type reports discussed next.

Component report

This report performs a project-wide extract of all components found on your wiring diagram set. This data includes component tags, location codes, location reference, description text, ratings, catalog information, and block names.

Wire From/To report

If you marked components and/or terminals with location codes, you can make good use of this report. This report first extracts component, terminal, location code, and wire connection information from every drawing in the project set. Then it displays a location list dialog box where you can make your report's "from" and "to" location selections. All the location codes AutoCAD Electrical found on the drawing set are listed on each side of this dialog box. Location "(??)" is also included in the list if AutoCAD Electrical found any component or stand-alone terminal that did not have an assigned location code.

Component Wire List report

This report extracts the component wire connection data and displays it in a dialog box. Each entry shows a connection to a component, the wire number, component tag name, terminal pin number, component location code (if present), and the layer that the connected wire is on.

Connector Plug report

This report extracts plug/jack connection reports and optionally generates pin charts. Since each wire tied to each connector pin displays in the report, each pin has two entries. One entry is for the 'in' wire and the other is for the 'out' wire. To create a more useful report, select the PIN chart 'on' radio button at the bottom of the Report Generator dialog box. Make sure the Remove duplicated pin numbers toggle is checked and click OK. It reformats the report so each pin is listed only once.

PLC I/O Address and Description report

This report lists each PLC module and its beginning and ending I/O address numbers. It scans your drawing set and returns all individual I/O connection points it finds. It includes up to five lines of description text and the connected wire number for each I/O point.

PLC I/O Component Connection report

This report scans the selected drawings and returns information about any components connected to PLC I/O points. Data for the report starts at each wire connection point and follows the connected wire. The first terminal symbol, fuse symbol, or connector symbol that is hit reports in the column marked with default "TERMTAG" label. The first schematic component reports in the column with the "CMPTAG" label. If there are multiple instances, the one closest to the PLC I/O point is the one that shows up in the report column.

PLC Modules Used So Far report

For this report, AutoCAD Electrical quickly scans the wiring diagram set. It returns in a few moments and displays the I/O modules it finds. Each entry shows the beginning and ending address of the module.

Terminal Numbers report

This project-wide, stand-alone report lists all instances of terminals. Each entry includes information tied directly to the terminal such as terminal number, terminal strip number, and location code.

Terminal Plan report

This project-wide, stand-alone report does a wire network extraction. It takes longer to generate, but the report includes wire number and wire layer name information.

Connector Summary report

This report lists a single line for each connector tag along with pins used, maximum pins allowed (if the parent carries the PINLIST information), and a list of any repeated pin numbers used. You can run this report across the project or for a single component.

Connector Details report

This report extracts the component wire connection data and displays it in a dialog box. Each entry shows a connection to a component, the wire number, component tag name, terminal pin number, component location code (if present), and the layer that the connected wire is on.

Cable Summary report

This project-wide cable conductor report gives a report listing all the cable marker tags (parent tags) found.

Cable From/To report

This project-wide cable conductor report lists the "from / to" for each cable conductor. It also lists the parent cable number of the conductor, conductor color code, and wire number (if present).

Wire Label report

This report lists wire markers/labels and can be used to create physical wire or cable labels.

MISSING CATALOG DATA

The **Missing Catalog Data** tool is used to display a diamond on the components in the schematic diagram and panel layout that do not have catalog data attached. The procedure to use this tool is given next.

- Click on the **Missing Catalog Data** tool from the **Schematic** panel. The **Show Missing Catalog Assignments** dialog box will be displayed; refer to Figure-13.

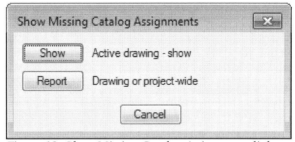

Figure-13. Show Missing Catalog Assignments dialog box

- Click on the **Show** button to highlighted the components not having catalog data attached; refer to Figure-14.

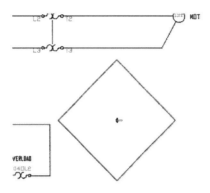

Figure-14. Component not having catalog data attached

- Click on the **Report** button from the **Show Missing Catalog Assignments** dialog box to generate the report. The **Schematic Reports** dialog box will be displayed.
- Click on the **OK** button from the dialog box. The **Select Drawings to Process** dialog box will be displayed.
- Select the **Do All** button to include all the drawings of the Project and click on the **OK** button from the dialog box. The **QSAVE** dialog box will be displayed.
- Click on the **OK** button from the dialog box. The **Report Generator** dialog box will be displayed with the missing entries; refer to Figure-15.

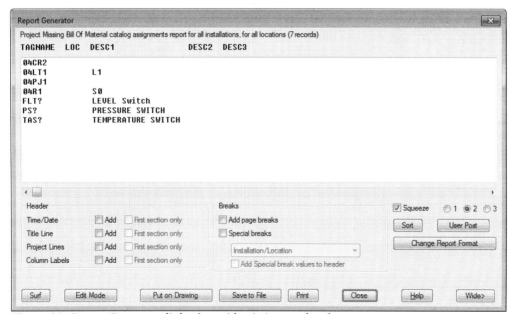

Figure-15. Report Generator dialog box with missing catalog data components

- Click on the **Put on Drawing** button or **Save to File** button as discussed earlier to extract the report. You can also print the table by using the **Print** button from the dialog box.

ELECTRICAL AUDIT

The **Electrical Audit** tool is used to audit all the drawings in project and find out the faulty components in the project. The procedure to use this tool is given next.

- Click on the **Electrical Audit** tool from the **Schematic** panel in the **Ribbon**. The **Electrical Audit** dialog box will be displayed.
- Click on the **Details** button. The expanded dialog box will be displayed; refer to Figure-16.

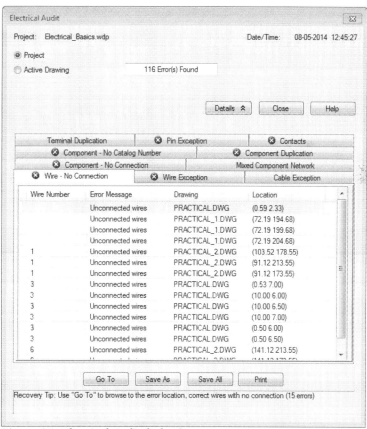

Figure-16. Electrical Audit dialog box

- Click on the Active Drawing radio button to display the problems in the current drawing. Click on the tabs in the expanded dialog box which are marked with a cross, the errors will be displayed; refer to Figure-17.

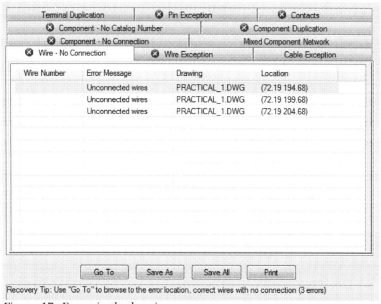
Figure-17. Errors in the drawing

- Select the error from the table and click on the **Go To** button from the dialog box. The component having error will be highlighted in the drawing; refer to Figure-18.

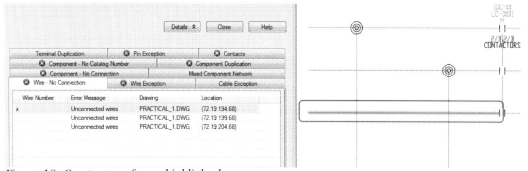
Figure-18. Component of error highlighted

- Modify the component as per the requirement.

DRAWING AUDIT

The **DWG Audit** tool is used to audit the drawing for errors of wiring or wire numbers. The procedure to use this tool is given next.

- Click on the **DWG Audit** tool from the **Schematic** panel of the **Ribbon**. The **Drawing Audit** dialog box will be displayed; refer to Figure-19.

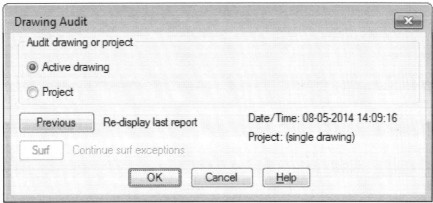

Figure-19. Drawing Audit dialog box

- Select the **Active drawing** or **Project** radio button as per your requirement. If selected the Project radio button then you need to include the drawings of project as discussed earlier. (Active drawing radio button is selected in our case)
- Click on the **OK** button from the dialog box. The **Drawing Audit** dialog box with the options to display in the drawing will be displayed; refer to Figure-20.

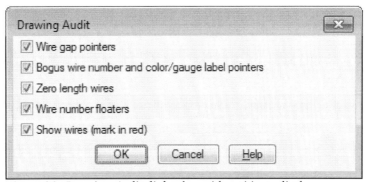

Figure-20. Drawing Audit dialog box with entities to display

- Select the desired check boxes and click on the **OK** button from the dialog box. The selected components will be highlighted; refer to Figure-21.

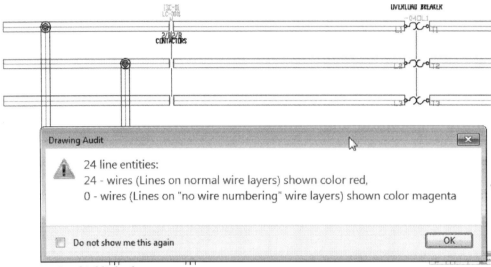

Figure-21. Wires highlighted

- Click on the **OK** button from the dialog box to exit.

Similarly, you can use the other tools in the Reports tab to edit or create the reports.

DYNAMIC EDITING OF REPORTS IN DRAWING

You can edit any entry of the reports inserted in the drawing by using the dynamic editing facility of the AutoCAD Electrical. The procedure is given next.

- Click in any cell of the table. The **Table Cell** tab will be displayed; refer to Figure-22.

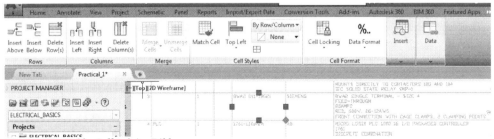

Figure-22. Table Cell tab in Ribbon

- Type the new value for the selected cell if you want to change the value.
- Using the tools in the **Table Cell** tab you can change the format of the table.

The tools in the **Table Cell** tab are discussed next.

MODIFYING TABLES

Select any cell in the table; the **Table Cell** tab will be added to the **Ribbon**, as shown in Figure-23. The options in the **Table Cell** tab are used to modify the table, insert block, add formulas, and perform other operations.

*Figure-23. The **Table Cell** tab added to the **Ribbon***

Modifying Rows

To insert a row above a cell, select a cell and choose the **Insert Above** tool from the **Rows** panel in the **Table Cell** tab. To insert a row below a cell, select a cell and choose the Insert Below tool from the Rows panel in the **Table Cell** tab. To delete the selected row, choose the **Delete Row(s)** tool from the **Rows** panel in the **Table Cell** tab. You can also add more than one row by selecting more than one row in the table.

Modifying Columns

This option is used to modify columns. To add a column to the left of a cell, select the cell and choose the **Insert Left** tool from the **Columns** panel in the **Table Cell** tab. To add a column to the right of a cell, select a cell and choose the **Insert Right** tool from the **Columns** panel in the **Table Cell** tab. To delete the selected column, choose the **Delete Column(s)** tool from the **Columns** panel in the **Table Cell** tab. You can also add more than one column by selecting more than one row in the table.

Merge Cells

This button is used to merge cells. Choose this button; the **Merge Cells** drop-down is displayed. There are three options available in the drop-down. Select multiple cells using the **SHIFT** key and then choose **Merge All** from the drop-down to merge all the selected cells. To merge all the cells in the row of the selected cell, choose the **Merge By Row** button from the drop-down. Similarly, to merge all cells in the column of the selected cell, select the **Merge By Column** tool from the drop-down. You can also divide the merged cells by choosing the Unmerge Cells button from the Merge panel.

Match Cells

This button is used to inherit the properties of one cell into the other. For example, if you have specified **Top Left Cell Alignment** in the source cell, then using the **Match Cells** button, you can inherit this property to the destination cell. This option is useful if you have assigned a number of properties to one cell, and you want to inherit these properties in some specified number of cells. Choose **Match Cell** from the **Cell Styles** panel in the **Table Cell** tab; the cursor is changed to the match properties cursor and you are prompted to choose the destination cell. Choose the cells to which you want the properties to be inherited and then press ENTER.

Table Cell Styles

This drop-down list displays the preexisting cell styles or options to modify the existing ones. Select the desired cell style to be assigned to the selected cell. The **Cell Styles** drop-down also has the options to create a new cell style or manage the existing ones.

Edit Borders

Choose the **Edit Borders** button from the **Table Cell** tab; the **Cell Border Properties** dialog box will be displayed, as shown in Figure-24.

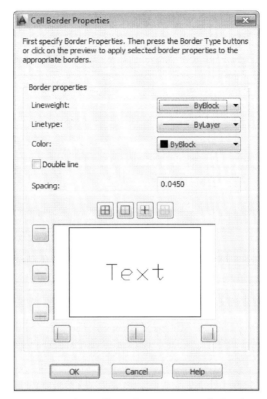

Figure-24. The Cell Border Properties dialog box

Text Alignment

The down arrow with the **Middle-Center** button of the **Cell Styles** panel in the **Ribbon** is used to align the text written in cells with respect to the cell boundary. Choose this button; the **Text Alignment** drop-down will be displayed. Select the desired text alignment from this drop-down; the text of the selected cell will get aligned accordingly.

Locking

This button is used to lock the cells so that they cannot be edited by accident. Select a cell and choose **Cell Locking** from the **Cell Format** panel in the **Table Cell** tab; the **Cell Locking** drop-down list will be displayed. Four options are available in the drop-down list. The **Unlocked** option is chosen by default. Choose the **Content Locked** option to prevent the modification in the content of the text, but you can modify the formatting of the text. Choose the **Format Locked** option to prevent the modification in the formatting of the text, but in this case, you can modify the content of the text. Select the **Content** and **Format Locked** option to prevent the modification of both formatting and content of the text.

Data Format

The display of the text in the cell depends on the format type selected. Choose **Cell Format** to change the format of the text in the cell. On doing so, the **Data Format** drop-down list is displayed. Choose the required format from it. You can also select the **Custom Table Cell Format** option and choose the required format from the Data type list box in the **Table Cell Format** dialog box and then choose the **OK** button.

Block

This tool is used to insert a block in the selected cell. Choose the **Block** tool from the **Insert** panel; the **Insert a Block in a Table Cell** dialog box will be displayed, refer to Figure-25. Enter the name of the block in the **Name** edit box or choose the **Browse** button to locate the destination file of the block. If you have browsed the file path, it will be displayed in the **Path** area. The options available in the **Insert a Block in a Table Cell** dialog box are discussed next.

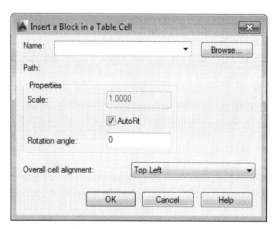

Figure-25. The **Insert a Block in a Table Cell** dialog box

Properties Area
The options in this area are discussed next.

Scale. This edit box is used to specify the scale of the block. By default, this edit box is not available because the **AutoFit** check box is selected below this drop-down list. Selecting the **AutoFit** check box ensures that the block is scaled such that it fits in the selected cell.

Rotation angle. The **Rotation angle** edit box is used to specify the angle by which the block will be rotated before being placed in the cell.

Overall cell alignment. This drop-down list is used to define the block alignment in the selected cell.

Field
You can also insert a field in the cell. The field contains the data that is associative to the property that defines the field. For example, you can insert a field that has the name of the author of the current drawing. If you have already defined the author of the current drawing in the Drawing Properties dialog box, it will automatically be displayed in the field. If you modify the author name and update the field, the changes will automatically be made in the text. When you choose the **Field** button from the **Insert** panel, the **Field** dialog box will be displayed. You can select the field to be added from the **Field names list** box and select the format of the field from the **Format list box**. Choose **OK** after selecting the field and format. If the data in the selected field is already defined, it will be displayed in the Text window. If not, the field will display dashes (----).

Formula
Choose **Formula** from the **Insert** panel; a drop-down list is displayed. This drop-down list contains the formulas that can be applied to a given cell. The formula calculates the values for that cell using the values of other cell. In a table, the columns are named with letters (like A, B, C, ...) and rows are named with numbers (like 1, 2,

3, ...). The TABLEINDICATOR system variable controls the display of column letters and row numbers. By default, the TABLEINDICATOR system variable is set to 1, which means the row numbers and column letters will be displayed when Text Editor is invoked. Set the system variable to zero to turn off the visibility of row numbers and column letters. The nomenclature of cells is done using the column letters and row numbers. For example, the cell corresponding to column A and row 2 is A2. For a better understanding, some of the cells have been labeled accordingly in Figure-26.

Formulas are defined by the range of cells. The range of cells is specified by specifying the name of first and the last cell of the range, separated by a colon (:). The range takes all the cells falling between specified cells. For example, if you write A2 : C3, this means all the cells falling in 2nd and 3rd rows, Column A and B will be taken into account. To insert a formula, double click on the cell; Text Editor is invoked. You can now write the syntax of the formula in the cell. The syntax for different formulas are discussed later while explaining different formulas. Formulas can also be inserted by using the Formula drop-down list. Different formulas available in the Formula drop-down list are discussed next.

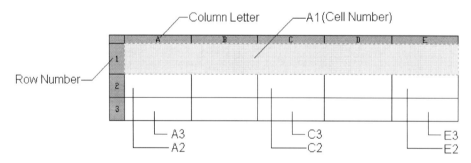

Figure-26. The Table showing nomenclature for Columns, Rows, and Cells

Sum

The **Sum** option gives output for a given cell as the sum of the numerical values entered in a specified range of cells. Choose the **Sum** option from the **Table Cell > Insert > Formula** drop-down; you will be prompted to select the first corner of the table cell range and then the second corner.

The sum of values of all the cells that fall between the selected range will be displayed as the output. As soon as you specify the second corner, the Text Editor is displayed and also the formula is displayed in the cell. In addition to the formula, you can also write multiline text in the cell. Choose **Close Text Editor** from the **Close** panel to exit the editor. When you exit the text editor, the formula is replaced by a hash (#). Now, if you enter numerical values in the cells included in the range, the hash (#) is replaced according to the addition of those numerical values. The prompt sequence, when you select the Sum option, is given next.

Select first corner of table cell range: Specify a point in the first cell of the cell range.

Select second corner of table cell range: Specify a point in the last cell of the cell range.

Note that the syntax for the Sum option is: =Sum{Number of the first cell of cell range (for example: A2): Number of the last cell of the cell range (for example: C5)}

Average

This option is used to insert a formula that calculates the average of values of the cells falling in the cell range. Prompt sequence is the same as for the Sum option.

Note that the syntax for the **Average** option is: =Average{Number of the first cell of the cell range (for example: A2): Number of the last cell of the cell range (for example: C5)}

Count

This option is used to insert a formula that calculates the number of cells falling under the cell range. The prompt sequence is the same as for the Sum option.

Note that the syntax for the Count option is: =Count{Number of the first cell of cell range (for example: A2): Number of the last cell of cell range (for example: C5)}

Cell

This option equates the current cell with a selected cell. Whenever there is a change in the value of the selected cell, the change is automatically updated in the other cell. To do so, choose the **Cell** option from **Table Cell > Insert > Formula** drop-down; you will be prompted to select a table cell. Select the cell with which you want to equate the current cell. The prompt sequence for the Cell option is given next.

Select table cell: Select a cell to equate with the current cell.

Note that the syntax for the **Cell** option is: =Number of the cell.

Equation

Using this option, you can manually write equations. The syntax for writing the equations should be the same as explained earlier.

Manage Cell Content

The **Manage Cell Content** button is used to control the sequence of the display of blocks in the cell, if there are more than one block in a cell. Choose **Manage Cell Contents** from the **Insert** panel; the **Manage Cell Content** dialog box will be displayed, see Figure-27. The options in the **Manage Cell Content** dialog box are discussed next.

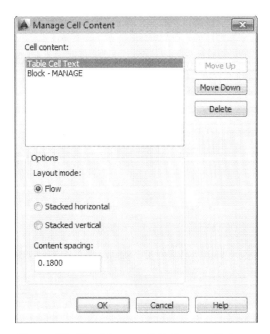

Figure-27. The Manage Cell Content dialog box

Cell content Area

This area lists all blocks entered in the cell according to the order of their insertion sequence.

Move Up

This button is used to move the selected block to one level up in the display order.

Move Down

This button is used to move the selected block to one level down in the display order.

Delete

This button is used to delete the selected block from the table cell.

Options Area

The options in this area are used to control the direction in which the inserted block is placed in the cell. If you select the **Flow** radio button, the direction of placement of the blocks in the cell will depend on the width of the cell. Select the **Stacked horizontal** radio button to place the inserted blocks horizontally. Similarly, select the **Stacked vertical** radio button to place the inserted blocks vertically. You can also specify the gap to be maintained between the two consecutive blocks by entering the desired gap value in the **Content** spacing edit box.

Link Cell

To insert data from an excel into the selected table, choose **Data > Link Cell** from the **Table Cell** tab; the **Select a Data Link** dialog box will be displayed. The options in this dialog box are used to link the data in a cell to the external sources.

Download from source

If the contents of the attached excel spreadsheet are changed after linking it to a cell, you can update these changes in the table by choosing this button. AutoCAD will inform you about the changes in the content of the attached excel sheet by displaying an information bubble at the lower-right corner of the screen.

Project

Chapter 11

```
The major topics covered in this chapter are:
```

- **Schematic Drawing**
- **Panel Layout**
- **Report Generation**

Project

Create schematic of the circuit as shown in Figure-1. Design the panel for the user and then generate the report for the components.

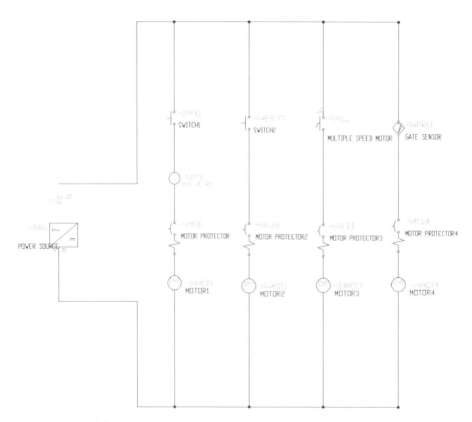

Figure-1. Project-Model

Starting a New Project File

- Click on the AutoCAD Electrical icon from the desktop or Start AutoCAD Electrical by using the Start menu.
- Click on the **New Project** button from the **Project Manager**; refer to Figure-2.

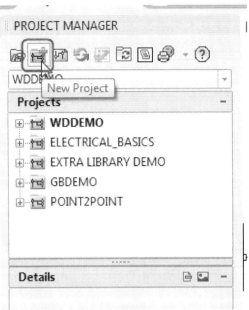

Figure-2. New Project button

- On doing so, the **Create New Project** dialog box will be displayed; refer to Figure-3.

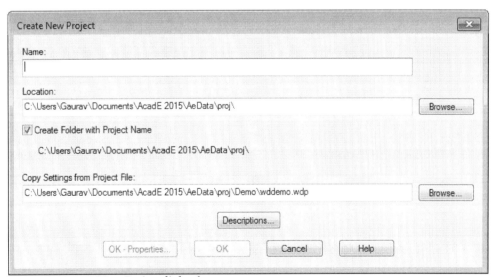

Figure-3. Create New Project dialog box

- Specify the name of the file as **Project** in the **Name** field of the dialog box.
- Click on the **OK-Properties** button from the dialog box. The **Project Properties** dialog box will be displayed.
- Click on the Drawing Properties tab from the dialog box.

The options in the dialog box will be displayed as shown in Figure-4.

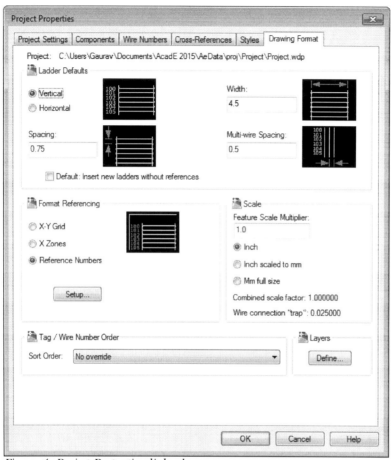

Figure-4. Project Properties dialog box

- Select the **Horizontal** radio button from the **Ladder Defaults** area of the dialog box and click on the **OK** button. A new project file will be added in the **Project Manager** with the name **Project**.

Adding Drawing in the Project

- Click on the **New Drawing** button from the Project Manager. The **Create New Drawing** dialog box will be displayed; refer to Figure-5.

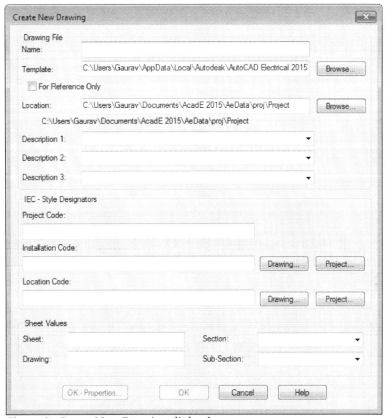

Figure-5. Create New Drawing dialog box

- Click in the **Name** field of the dialog box and specify **Project_1** as the name.
- Click on the **Browse** button for the **Template** edit box and select the **ACAD_ELECTRICAL_IEC** template from the dialog box displayed; refer to Figure-6.

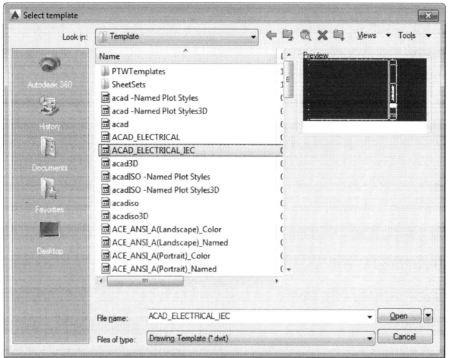

Figure-6. Select Template dialog box

- Click on the **Open** button from the dialog box.
- Click on the **OK** button from the Create New Drawing dialog box. The **Apply Project Defaults to Drawing Settings** dialog box will be displayed; refer to Figure-7.

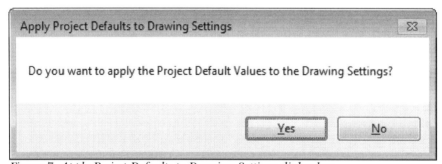

Figure-7. Apply Project Defaults to Drawing Settings dialog box

- Click on the **No** button from the dialog box. New drawing will be added in the Project with the name Project_1.
- Click on the **Grid Snap** and **Grid Display** button to turn off the grid snap and hide the grid from drawing.

Creating Ladder

- Click on the **Insert Ladder** button from the **Ladders** drop-down in the **Insert Wires/Wire Numbers** panel. The **Insert Ladder** dialog box will be displayed; refer to Figure-8.

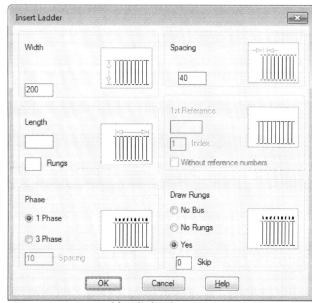

Figure-8. Insert Ladder dialog box

- Click on the **OK** button from the dialog box. You are asked to specify the start position of the first rung.
- Click in the drawing area to specify the starting position. The end point of the ladder gets attached to the cursor and you are prompted to specify the last rung.
- Click in the drawing area to specify the last rung so that the ladders are displayed as shown in Figure-9.

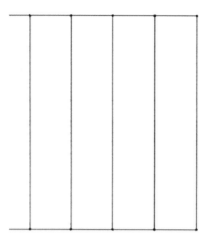

Figure-9. Ladders created

Adding wire

- Click on the **Wire** tool from the **Insert Wires/Wire Numbers** panel in the **Ribbon**. You are asked to specify the start point of the wire.
- Click on the top end of the ladder as shown in Figure-10.

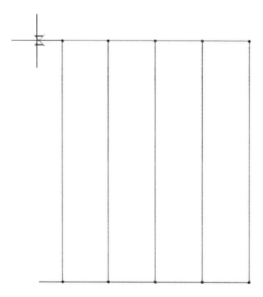

Figure-10. End to be selected

- Drawing the wire as shown in Figure-11.

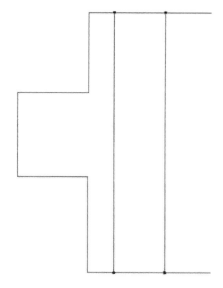

Figure-11. Wire drawn

Adding Components in the circuit

- Click on the **Icon Menu** button from the **Insert Components** panel in the **Ribbon**. The **Insert Component** dialog box will be displayed; refer to Figure-12.

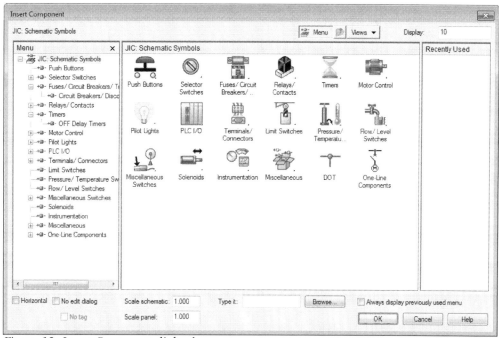

Figure-12. Insert Component dialog box

- Click on the **Miscellaneous** category from the dialog box. The components will be displayed as shown in Figure-13.

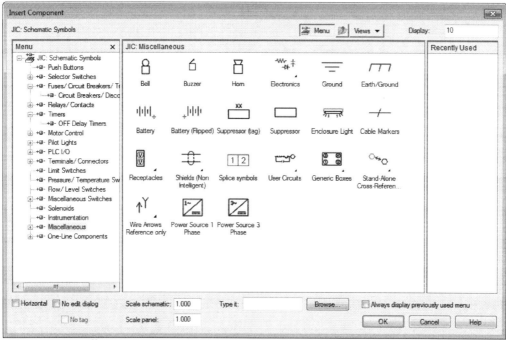

Figure-13. Miscellaneous components

- Click in the **Scale schematic** edit box at the bottom of the dialog box and specify the value as **20**; refer to Figure-14.

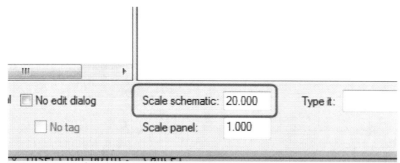

Figure-14. Scale schematic edit box

- Click on the **Power Source 1 Phase** component from the dialog box. The component will get attached to the cursor.
- Click in the middle of the wire to place the component; refer to Figure-15. The **Insert/Edit Component** dialog box will be displayed; refer to Figure-16.

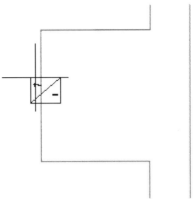

Figure-15. Placing power source

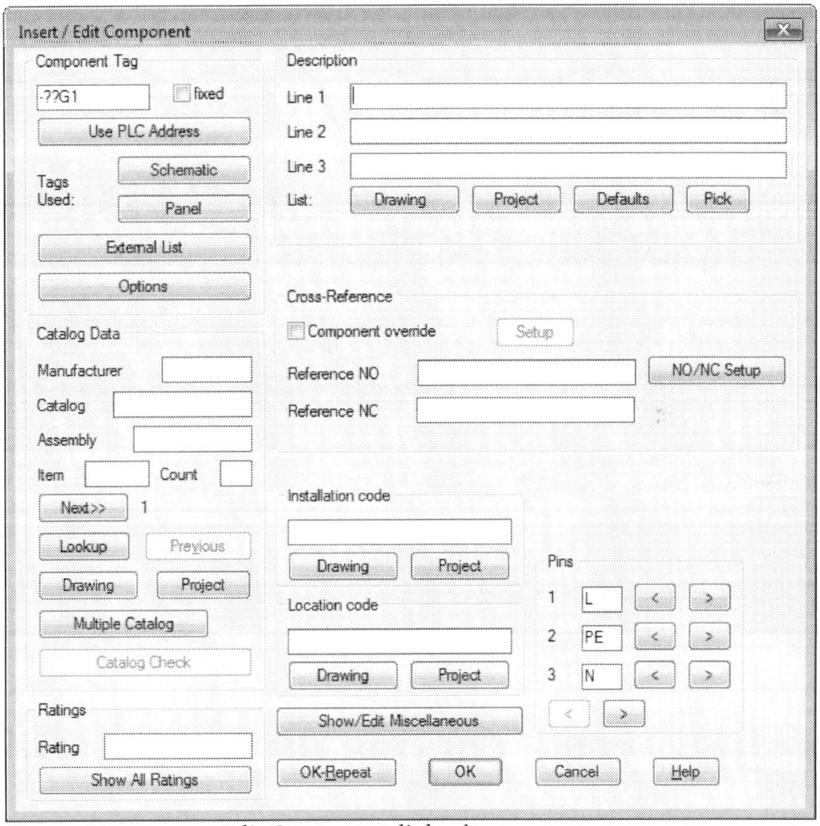

Figure-16. Insert or Edit Component dialog box

- Click in the **Line 1** edit box of the **Description** area of the dialog box and specify **Power Source** in it.
- Click on the **Lookup** button from the **Catalog Data** area of the dialog box. The **Catalog Browser** will be displayed; refer to Figure-17.

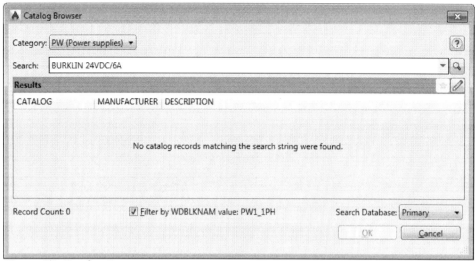

Figure-17. Catalog Browser

- Click in the **Search** field and specify **230VAC**.
- Click on the **Search** button next to **Search** field. The list of power sources will be displayed; refer to Figure-18.

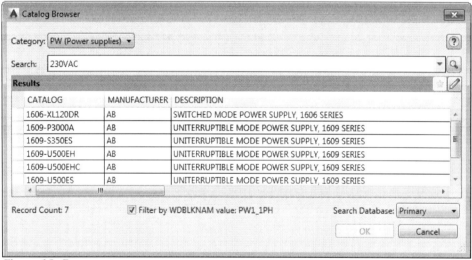

Figure-18. Power sources

- Select the **1609-P3000A** field under the **CATALOG** column and click on the **OK** button from the dialog box.
- Click on the **Next** button from the **Catalog Data** area of the dialog box to specify the item number.
- Specify the **Installation Code** and **Location Code** as **X1** and **L3** respectively.

- Click on the **OK** button from the dialog box. The power source will be placed; refer to Figure-19.

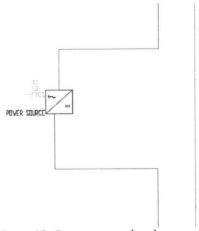

Figure-19. Power source placed

Adding other components

- Click on the Icon Menu button once again and Select the Motor Control category. The components for motor controls will be displayed; refer to Figure-20.

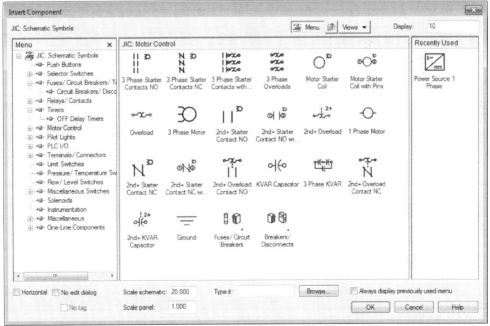

Figure-20. Components of Motor Control

- Click on the **1 Phase Motor** component. It will get attached to the cursor.
- Click on the first rung of the ladder and place the component as shown in Figure-21.

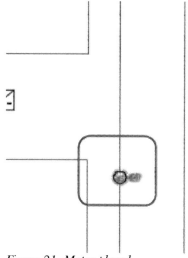

Figure-21. Motor placed

- Click in the **Line 1** edit box in the **Description** area of the dialog box displayed and specify the name as **Motor1**.
- Click on the **Lookup** button from the **Catalog Data** area of the dialog box and select the **1329-ZF00206NVH** field in the **CATALOG** column; refer to Figure-22.

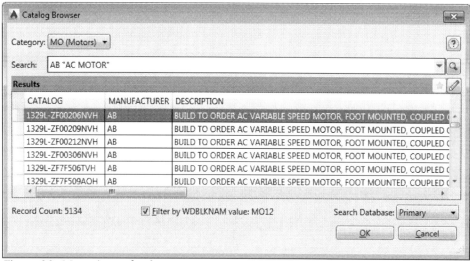

Figure-22. Motor in catalog browser

AutoCAD Electrical 2015 Black Book   11-15

- Click on the **OK** button from the **Catalog Browser** dialog box.
- Click on the **Next** button from the **Catalog Data** area of the dialog box to specify the item number.
- Specify the same Installation and Location codes as done for previous component.
- Click on the **OK-Repeat** button from the dialog box and create motors for other rungs.
- Similarly add the other components in the schematic drawing; refer to Figure-23.

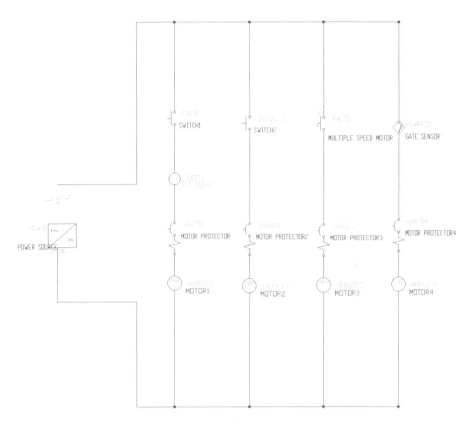

Figure-23. Project-Model

Adding Panel drawing in the Project

- Click on the **New Drawing** button from the **Project Manager**. The **Create New Drawing** dialog box will be displayed as discussed earlier.
- Specify the name as **Panel Drawing** in the **Name** field of the dialog box.

- Specify the same Installation and Location codes, and click on the **OK** button from the dialog box.
- Click on No button from the dialog box displayed for applying defaults of project.

Adding panel components

- Click on the **Panel** tab from the Ribbon. The tools related to panel will be displayed.
- Click on the **Icon Menu** button from the **Insert Component Footprints** panel. The **Alert** dialog box will be displayed.
- Click on the **OK** button from the dialog box. The **Insert Footprint** dialog box will be displayed; refer to Figure-24.

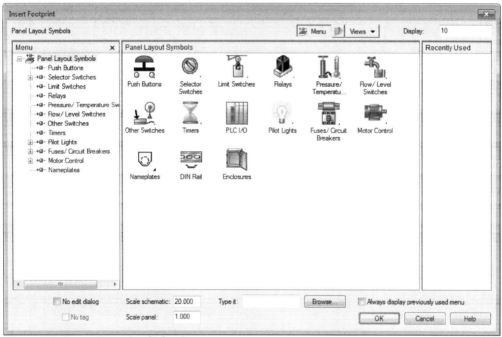

Figure-24. Insert Footprint dialog box

- Click on the **Enclosures** category from the dialog box. The **Footprint** dialog box will be displayed as shown in Figure-25.

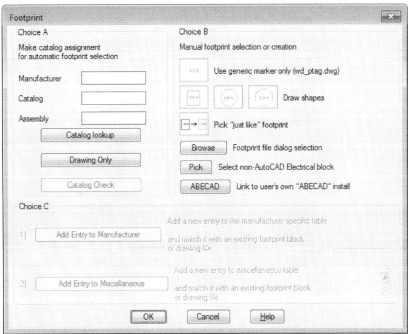

Figure-25. Footprint dialog box

- Click on the **Catalog lookup** button from the dialog box.
- Make the **Search** field empty and click on the **Search** button from the **Catalog Browser**. The list of enclosures will be displayed; refer to Figure-26.

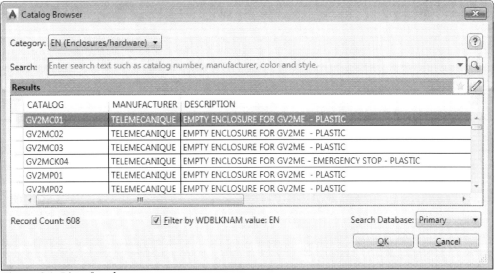

Figure-26. List of enclosures

- Select the first enclosure from the list and click on the **OK** button from the **Catalog Browser**.
- Click on the **Browse** button from the **Footprint** dialog box. The **Pick** dialog box will be displayed; refer to Figure-27.

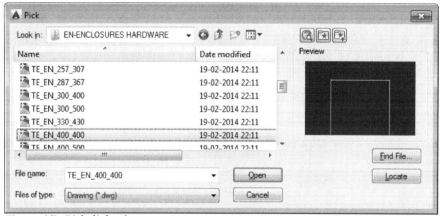

Figure-27. Pick dialog box

- Select the desired file and click on the **Open** button from the dialog box. The selected enclosure will get attached to the cursor.
- Click in the drawing area to place the enclosure. You are asked to specify the rotation angle for the enclosure.
- Specify the rotation as **0** and press **ENTER**. The **Panel Layout-Component Insert/Edit** dialog box will be displayed; refer to Figure-28.
- Click in the **Line 1** of **Description** area and specify the value as **Panel** enclosure.
- Click on the **Next** button from the **Item Number** area to specify the item number.
- Click on the **OK** button from the dialog box.

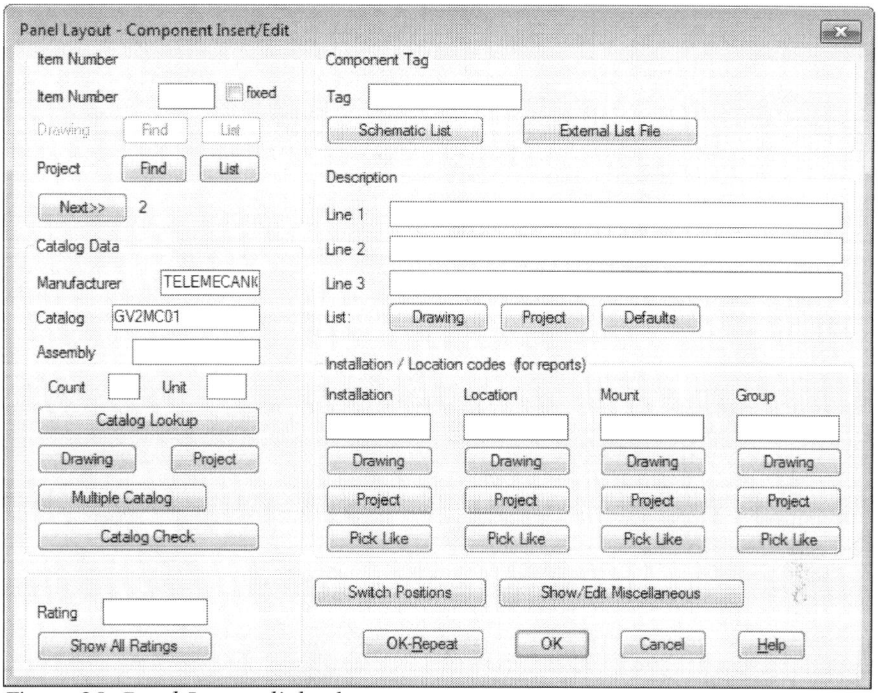

Figure-28. Panel Layout dialog box

Inserting buttons in the Panel layout

- Click on the **Schematic List** button from the **Insert Component Footprints** drop-down in the **Insert Component Footprints** panel in the **Ribbon**. The **Schematic Components List** dialog box will be displayed as shown in Figure-29.

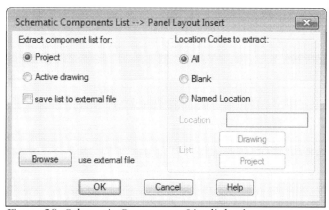

Figure-29. Schematic Components List dialog box

- Click on the **OK** button from the dialog box. The **Select Drawings to Process** dialog box will be displayed; refer to Figure-30.

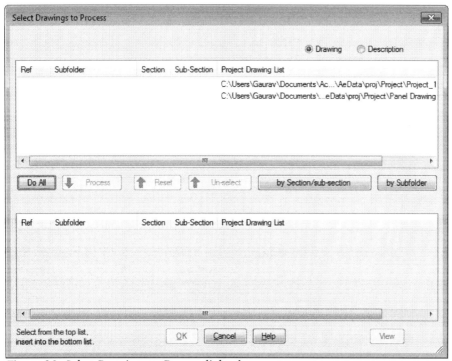
Figure-30. Select Drawings to Process dialog box

- Select the **Project_1** drawing from the list and click on the **Process** button from the dialog box.
- Click on the **OK** button from the dialog box. The **Schematic Components (active project)** dialog box will be displayed; refer to Figure-31.

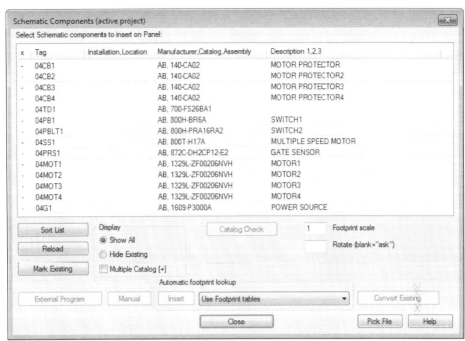

Figure-31. Schematic Components dialog box

- Select the component from the list for which you want to insert the footprint and click on the **Insert** button from the dialog box.
- If the footprint is mapped to component then the footprint will get attached to the cursor and you are asked to place the footprint.
- Place the footprint in the panel enclosure. The **Panel Layout** dialog box will be displayed. Specify the desired parameters and click on the **OK** button from the dialog box.
- If the footprint for the component is not mapped then the **Footprint** dialog box will be displayed and you need to manually select the block for the component by using the **Browse** button as done earlier.
- Similarly, place all the desired components in the panel enclosure. After adding all the components, the panel is displayed as shown in Figure-32.

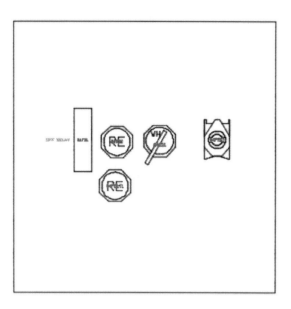

Figure-32. Panel after adding components

- Specify the desired annotation to the buttons to make it easy for interpretation.

Generating Reports

- Click on the **Reports** tab in the **Ribbon**. The tools related to reports will be displayed; refer to Figure-33.

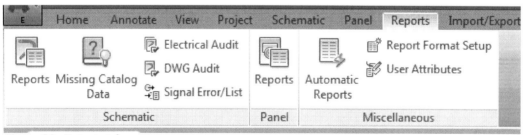

Figure-33. Reports tab

- Click on the **Reports** tool from the **Schematic** panel in the tab. The **Schematic Reports** dialog box will be displayed; refer to Figure-34.

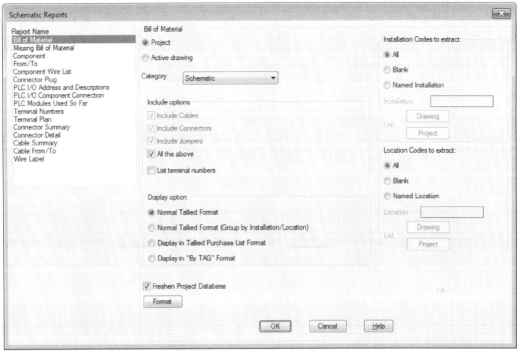

Figure-34. Schematic Reports dialog box

- Select the **Bill of Material** option from the **Report Name** selection box.
- Click on the **OK** button from the dialog box. The **Select Drawings to Process** dialog box will be displayed.
- Click on the **Do All** button and click on the **OK** button from the dialog box. The **QSAVE** dialog box will be displayed if the drawing is not saved.
- Click on the **OK** button from the dialog box. The **Report Generator** dialog box will be displayed; refer to Figure-35.

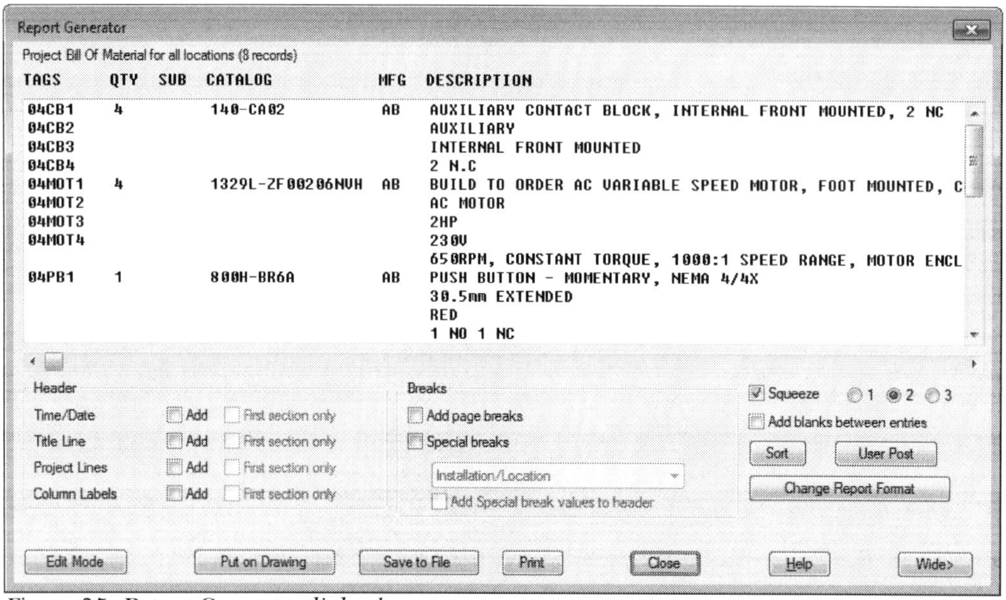

Figure-35. Report Generator dialog box

- Click on the **Put on Drawing** button. The **Table Generation Setup** dialog box will be displayed; refer to Figure-36.

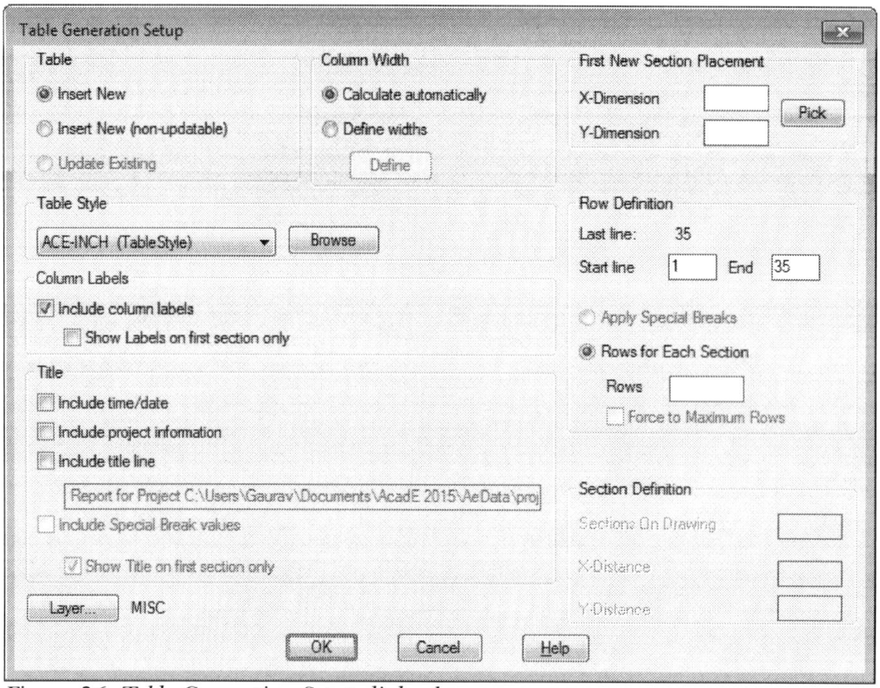

Figure-36. Table Generation Setup dialog box

- Click on the **OK** button from the dialog box. The table will get attached to the cursor.
- Click at the desired location in the drawing area to place the table. Refer to Figure-37.

Figure-37. Report

- Click on the **Close** button from the dialog box to exit.
- Similarly, you can insert other reports as required.

Page left blank for Student Notes

Index

Symbols

3 Phase tool 5-19
3 Phase Wire Numbering dialog box 5-19
22.5 Degree 5-4
45 Degree 5-4
67.5 Degree tool 5-4
.wde file 4-22

A

ACAD Electrical.dwt 2-10
ACAD Electrical IEC.dwt 2-10
Active Drawing button 6-23
Add from Catalog button 10-7
Additional angles check box 2-45
Add New USER data record dialog box 6-11
Add Rung tool 6-35
Align tool 6-17
Allow spacers/breaks radio button 7-13
Annotation buttons 2-46
Another Bus (Multiple Wires) radio button 5-8
apped wire pointer problem dialog box 6-14
Application button 2-13
Application Menu 2-13
Applying Password 2-28
Archive option 2-33
Assigned Jumpers area 6-6
Assign Wire Numbering Formats by Wire Layers dialog box 5-18
Associate Terminals tool 7-24
Attribute Size option 8-22
AutoCAD Electrical-Publish Setup dialog box 3-25
AutoCAD Electrical Publish to Web dialog box 3-23
Autodesk 360 button 2-11
Autodesk - Sign In dialog box 2-11

B

Balloon tool 9-14
Batch Plotting Options 3-20
Bend Wire tool 6-39
Bill of Material reports 10-4
Bill of Materials option 10-4
Block Diagram 1-5
Browse For Folder dialog box 3-5, 3-23

Build Up or Down? dialog box 8-11
Bump - Up/Down by radio button 3-17

C

Calculate Next button 4-6
Capacitors 1-10
Catalog Browser 4-14
Catalog Browser dialog box 4-8
Catalog Data area 4-7
Catalog edit box 4-7
Catalog Lookup File dialog box 3-10
Catalog Lookup File Preference area 3-9
Category drop-down 4-8, 10-4
Cell Border Properties dialog box 10-21
Cell Styles drop-down 10-20
Change Attribute Size dialog box 8-22
Change/Convert Wire Type tool 6-45
Check/Trace Wire tool 6-42
Circuit Builder tool 5-31
Circuit Diagram 1-3
Clean Screen button 2-47
Colors button 2-15
Command Window 2-39
Component Cross-Reference Display area 4-12
Component Cross-Reference tool 6-28
Component Cross-Reference Update check box 3-16
Component (Multiple Wires) radio button 5-7
Component override check box 4-11
Component Retag check box 3-16
Component Tag area 4-4, 4-6
Conductors 1-6
Confirm Password dialog box 2-29
Connected PLC Address Not Found dialog box 4-5
Connectors and terminals 1-8
Copy Assembly tool 9-31
Copy Catalog Assignment tool 6-8
Copy Codes drop-down 9-30
Copy Component tool 6-13
Copy Footprint tool 9-27
Copy Location Code tool 9-30
Copy Settings from Project File edit box 3-5
Copy Tag or Calculate Next button 4-6
Count edit box 4-7
Create/Edit Wire Type tool 6-43
Create Folder with Project Name check box 3-5
Create New Drawing dialog box 3-12
Create New Project dialog box 3-4

Crosshairs option 2-16
Cross-Reference Area 4-11
Cross-Reference Component Override dialog box 4-11
Cross-Reference Report dialog box 6-29
Customization button 2-47

D

Dark drop-down 2-14
Default templates 2-8
Defined Terminal Wiring Constraints Exceeded dialog box 9-19
Delete Footprint tool 9-27
Delete Wire Numbers tool 6-35
Detailed Plot Configuration mode button 3-20
Detailed Plot Configuration Option dialog box 3-21
Diodes and rectifiers 1-14
Display grid beyond Limits check box 2-42
Display in By TAG Format 10-5
Display in Tallied Purchase List Format 10-5
Display Resolution 2-3
Download from source 10-28
Download Offline Help option 2-12
Drafting Settings dialog box 2-42, 2-44
Drawing Area 2-37
Drawing Audit dialog box 6-41, 10-17
Drawing button 7-16
Drawing List Display Configuration button 3-18
Drawing option 2-18
Drawing Properties dialog box 5-14
Drawing Properties option 5-14
Drawing tab bar 2-36
Drawing Template (*.dwt) option 2-19
Drawing Window Colors dialog box 2-15
DWG Audit tool 10-17

E

Earthing 1-15
Edit Components panel 6-2
Editor tool 9-18
Edit tool 6-2
Edit Wire Number tool 6-31
Edit Wires/Wire Numbers panel 6-31
elect Project File dialog box 3-11
Electrical Audit dialog box 10-15
Electrical Audit tool 10-15
Electrical Code Standard drop-down 3-11
Electrical Templates 2-10
Email option 2-33

Empty Space 5-9
Encrypt drawing properties 2-29
Equipment List Spread Settings dialog box 4-22
Equipment List tool 4-20
Error/Exception Report dialog box 6-28
eTransmit option 2-33
Excel spreadsheet format(.xls) 10-9
Exchange Apps button 2-12
Export Data dialog box 2-32
Export option 2-31

F

Files of type drop-down 2-27
Find/Replace tool 6-33
Fit to paper check box 2-34
Fix tool 6-32
Fix/UnFix Component Tag dialog box 6-8
Fix/UnFix Tag tool 6-8
Flip Wire Number tool 6-46
Format button 10-5
Formula 10-23
Fuses 1-11

G

Gap tool 5-6
Go Horizontal radio button 5-9
Go Vertical radio button 5-9
Grid Display button 2-40
Grid Snap button 2-40

H

Hide/Unhide Cross-Referencing tool 6-29
Hydraulic Components 4-27

I

Icon Menu button 4-2
Icon Menu tool 9-4
identifiers 4-12
IEC - Style Designators area 3-14
IGES (*.iges) option 2-32
Image drop-down 3-23
Imperial 2-9
import DGN format 2-24
Include unused/extra connections check box 7-13
Increment angle drop-down 2-45
Inductors and transformers 1-9

In-Line Wire Labels tool 5-25
Insert a Block in a Table Cell dialog box 10-22
Insert Angled Tee Markers tool 5-30
Insert Component Footprints drop-down 9-4
Insert Component Footprints -- Manual dialog box 9-11
Insert Component window 4-2
Insert Connector dialog box 7-20
Insert Connector (From List) tool 7-22
Insert Connectors drop-down 7-19
Insert Connector tool 7-20
Insert Dot Tee Markers tool 5-29
Insert/Edit Cable Marker (Parent wire) dialog box 5-27
Insert/Edit Component dialog box 4-4, 4-18, 6-3
Insert Footprint dialog box 9-4
Insert Hydraulic Components tool 4-27
Insert Ladder tool 5-11
Insert P&ID Components tool 4-27
Insert PLC (Full Units) tool 7-13
Insert PLC (Parametric) tool 7-11
Insert Pneumatic Components button 4-26
Insert Right tool 10-19
Insert Splice tool 7-23
Insert Terminals (Schematic List) 9-25
Insert Wires/Wire Numbers panel 5-2
Installation Code 4-13
Interconnect Components tool 5-5
Internal Jumper 6-4
Internal Jumper tool 6-4
Isolate Objects button 2-47
Isometric Drafting button 2-45

L

Labeling 1-18
Ladder Diagrams 5-10
Ladder drop-down 5-10
Layout-Component Insert/Edit dialog box 9-8
Library Icon Menu Paths 3-9
Light option 2-14
Link Cell 10-28
Location Code 4-13

M

Make it Fixed check box 6-32
Manage Cell Content dialog box 10-27
Manufacturer edit box 4-7, 7-17
Manufacturer Menu tool 9-12
Map symbol to catalog number button 4-18

Match Cells button 10-20
Maximum NO/NC counts and/or allowed Pin numbers dialog box 4-13
Memory 2-3
Metric 2-10
Missing Catalog Data tool 10-13
Module Layout dialog box 7-12
Move Circuit tool 6-15
Move Component tool 6-18
Move Wire Number tool 6-35
Multiple Bill of Material Information dialog box 4-9
Multiple Bus tool 5-6
Multiple Cable Markers tool 5-29
Multiple Catalog button 4-8

N

NC 4-12
New Drawing button 3-12
New option 2-17
New Project button 3-4
NO 4-12
Normal Tallied Format 10-5
Normal Tallied Format (Group by Installation/Location) 10-5
Number of Terminal Blocks edit box 9-22

O

Offline Help option 2-12
OK Insert Child button 5-28
Open options 2-24
Open Project button 3-11
Open Read-Only option 2-26
Open with no Template - Imperial button 2-19
Optional Script File dialog box 10-9
Options button 2-14
Options dialog box 2-14
Option: Tag Format "Family" Override dialog box 7-17
Ortho Mode button 2-43
Other Formats button 2-32

P

Panel Assembly tool 9-17
Panel balloon setup dialog box 9-14
Panel Layout - Component Insert/Edit dialog box 9-26
Panel Layout-Component Insert/Edit dialog box 9-11
Panel List tool 4-24
Papersize area 3-21
Partial Open dialog box 2-26

Partial Open option 2-26
Partial Open Read-Only option 2-26
Parts list 1-6
Pick Master button 9-31
P&ID components 4-27
Pins area 4-14
PLC I/O Address and Description report 10-11
PLC I/O Point List (this drawing)dialog box 7-16
PLC I/O tool 5-20
PLC Parametric Selection dialog box 7-11
Plot dialog box 2-33
Plot option 2-33
Plot Project tool 3-20
Pneumatic components 4-25
Polar Tracking button 2-43
Printer/plotter 2-34
Print options 2-33
Project Description dialog box 3-5
Project Manager 3-4
Project Manager button 3-3
Project Properties 3-8
Project Properties dialog box 3-6
Project Task List button 3-15
Project Wide Update/Retag button 3-15
Publish - Save Sheet List dialog box 3-27
Publish to DWF/PDF/DWFx tool 3-25
Publish to WEB tool 3-22
Push Buttons 4-17
Put on Drawing button 10-7

Q

Quick Access Toolbar 2-10

R

Refresh button 3-15
Remove option 3-30
Rename option 3-30
Renumber Ladder Reference tool 6-37
Renumber Ladders dialog box 3-17
Replace option 3-30
Report Generator dialog box 10-7
Report Name selection box 10-3
Reports tab 10-2
Reports tool 10-2
report types 10-3
Resequence Item Numbers tool 9-29
Resequence radio button 3-17

Resistors 1-9
Retag Components tool 6-21
Reverse/Flip Component tool 6-20
Revise Ladder tool 6-36
Run Script button 10-9

S

Save 2-26
Save Circuit to Icon Menu dialog box 6-16
Save Circuit to Icon Menu tool 6-15
Save Drawing As dialog box 2-27
Save Report to File dialog box 10-8
Schematic button 4-6
Schematic Components (active drawing) dialog box 9-10
Schematic Components List dialog box 9-9
Schematic --> Layout Wire Connection Annotation dialog box 9-16
Schematic Libraries node 3-9
Schematic List button 9-8
Schematic List tool 9-9
Schematic or Panel radio button 4-19
Schematic Reports dialog box 10-2
Scooting 6-17
Scoot tool 6-17
Search Box 2-11
Search button 4-15
Search edit box 4-15
Search for / Surf to Children dialog box 9-28
Search for/ Surf to Children dialog box 6-12
Search in Help 2-38
Search on Internet button 2-38
Security Options 2-28
Select Drawings to Process dialog box 3-17, 3-20, 4-25
Selection Cycling button 2-42
Select Schematic component or circuit dialog box 4-20
Select template dialog box 3-13
Select Wire Layer dialog box 5-18
Send to 3D Print Service option 2-33
Sequential Code drop-down 4-9
Set Wire Type dialog box 5-3
Sheet 1: Insert Ladder dialog box 5-11
Sheet Set option 2-18
Sheet Set Properties button 2-23
Show all components for all families check box 4-6
Show Missing Catalog Assignments dialog box 10-13
Show Wire tool 6-40
Sign In button 2-11
Special Annotation area 5-32

Specifications of PLCs 7-10
Specify PDF File dialog box 3-26
Start Drawing button 2-7
Stretch Wire tool 6-39
Swap Block/Update Block/Library Swap dialog box 6-24
Swap tool 6-32
Swap/Update Block tool 6-22
SWG system 1-17
Switch contacts 1-12
Switch types 1-13
System requirements 2-2

T

Table Cell tab 10-19
Table Edit dialog box 4-21
Table Generation Setup dialog box 10-7
Table Generator tool 9-22
TABLEINDICATOR system variable 10-24
Tag edit box 7-16
Tag/Wire Number Sort Order drop-down 3-11
Template edit box 3-14
Terminal Strip Selection dialog box 9-19
Terminal Strip Table Settings dialog box 9-23
Title Bar 2-10
Toggle NO/NC tool 6-22
Toggle Wire Number In-Line tools 6-46
Tracking Settings button 2-44
Transform Component drop-down 6-17
Trim Wire tool 6-34
TRMS category 7-23
Type a Command box 2-38
Types of Wires 1-16

U

Units option 2-49
Update a Block radio button 6-23
Update Stand-Alone Cross-Referencing tool 6-29
Use PLC Address button 4-4
User Defined List tool 4-18
User Table Data tool 6-10

V

Vendor Menu Selection dialog box 9-12

W

Window Elements area 2-14

Wire Annotation tool 9-15
Wire Color/Gauge Labels tool 5-23
Wire Editing drop-down 6-38
Wire Jumpers dialog box 6-4
Wire label color/gauge setup dialog box 5-23
Wire Number Leader tool 5-22
Wire Numbers tool 5-17
Wire Tagging (Project-Wide) dialog box 3-16
Wire tool 5-2
wireType button 5-3
Wiring Diagram 1-4
Wiring Schedule 1-4
Workflow 3-2

X

X-Y Grid Setup dialog box 5-12
XY Grid Setup tool 5-12
X Zones Setup tool 5-13

Z

Zip Project tool 3-28